ARCHAEOLOGY
and the Book of Mormon

A Cultural Perspective

Ian Main

DIGITAL LEGEND

ISBN: 978-1-934537-88-6

Digital Legend Press and Publishing
Salt Lake City, UT, and Rochester, NY
www.digitalegend.com

Send inquiries to: info@digitalegend.com or call 877-222-1960

*Thanks to Doug and Latricia Booth and Bishop George
who were so generous and hospitable on my first trip to Ohio,
to my wife Katherine for her patience and support
and to Campbell Gray for his expertise and guidance.*

Contents

Preface

MEMBERS OF THE CHURCH of Jesus Christ of Latter-Day Saints have long wrestled with the questions, "Where did the events of the Book of Mormon take place?" and "Who are the descendants of those people?" Such strivings are only natural as the Book of Mormon is central to our beliefs and anything that supports its truthfulness and links it to the physical world we see today is reassuring and confirming.

For much of my life the prevailing focus has been on Central and South America. Some years ago, missionaries carried copies of *Ancient America Speaks*, a Church produced film that linked the Book of Mormon with the ancient inhabitants of that region. At the same time Books of Mormon included artwork by Arnold Friberg showing Mayan buildings in the background. The same artwork was used on Church produced manual covers and adorned chapel walls. Later, videos and films continued the Central American theme.

During this period the Church maintained that the Hill Cumorah was located in New York state, several thousand miles from Central America. A number of theories were proposed to account for the apparent discrepancy. Some suggested that there were two hills named Cumorah: One in New York and another, the site of the final conflict, in Central America. Some said that the Nephites retreated northward and were pursued by their enemies from Central America all the way to New York. Others conjectured that while much of the Book of Mormon took place in Central America, some people migrated far to the north.

In more recent times, the Central American theory has been challenged. Critics of the Church have used DNA studies in an attempt to discredit any linkage between Central American populations and Hebrew ancestry (DNA vs. The Book of Mormon, 2003). In response, proponents of the Central American theory have proffered "population bottleneck and genetic drift" as a possible explanation for why Central and South American populations lacked any trace of Middle Eastern DNA (Book of Mormon and DNA Studies, 2014). Population bottleneck and genetic drift requires the death of large numbers of people from a population followed by a gradual loss of genetic markers.

Other Church members have taken a different approach to Book of Mormon geography. Perhaps spurred by the DNA debate a growing band of Saints, led by people such as Rod Meldrum and Wayne May, have argued that North America is the actual setting for the Book of Mormon and have linked the Hopewell Culture to the Nephite race. Rather than stretching the geography from Central America to New York state they have proposed a setting from the Gulf Coast northward through the Mississippi Valley to the Great Lakes region (Meldrum, 2011).

The existence of a mid-west United States theory came as a surprise to me. As a member of the Church since the age of 16, I had spent most of my life picturing the Book of Mormon through an Arnold Friberg filter. I had no knowledge of the prehistoric period of North America and the Cultures that inhabited that land.

Not long after encountering the North American theory I accompanied my wife to Ohio for unrelated business. In the course of that trip, I took the opportunity to visit a number of sites of the Hopewell and Adena cultures and met with people that had been studying those cultures for many years. I am grateful for their hospitality and generosity.

In this book I will briefly revisit the arguments of Rod Meldrum that initially pointed me towards North American cultures. I will then outline some of the major cultural practices and traits that I have uncovered that may strengthen the argument for a North American setting. I hope you enjoy these writings.

1

A Forgotten Past

THE VAST MAJORITY OF PEOPLE today are ignorant of the prehistory of North America. We have all heard about the ancient civilisations of South America but, for most people, pre-Columbian North America remains a mystery that has been largely ignored. This situation is very different from the world of nineteenth century Americans.

When white settlers first ventured from the east coast and began their westward movement across the North American continent, they could never have imagined what lay ahead. As they penetrated the Appalachian Mountains into the Mississippi Valley, they were astounded to discover tens of thousands of ancient structures. Not buildings of stone carved with pagan motifs but earthen mounds, some of astonishing size and complexity (Silverberg, 1970, pp. 10-11). Who constructed these mounds and why, was a matter of great conjecture.

The fledgling Smithsonian Institute felt the question of the mounds so important it gambled its own credibility on its first publication *Ancient Monuments of the Mississippi Valley*, documenting many of the more significant finds. The book "set a scholarly and scientific precedent for the institution" and helped to "establish the Smithsonian's credibility" (Squier & Davis, 1988, p. 1).

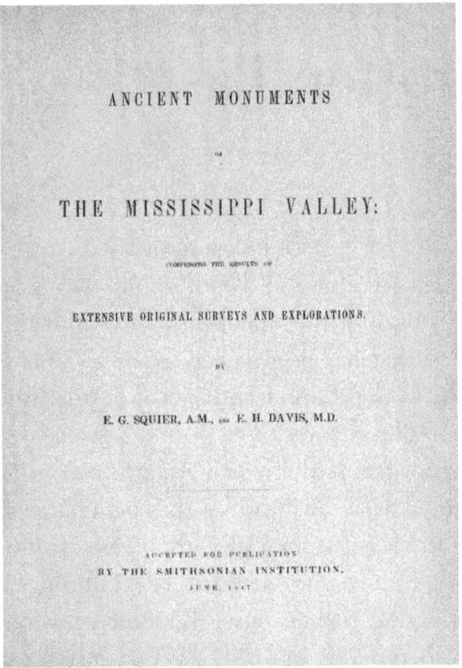

More information about the nature and purpose of the mounds will be provided in later chapters but, before pursuing that line of enquiry, let us focus on the impact these ancient monuments had on the people who saw them with their own eyes before many of those structures were damaged or destroyed by modern civilisation. These earthworks were so large, so numerous and some so technologically advanced that their construction by an ancient people was almost inconceivable. To add to the intrigue, some of the structures were clearly of a military nature. Who created these mounds and why became the subject of a debate that raged in North America through much of the nineteenth century (Silverberg, 1970, p. 15).

Many eyewitnesses could not believe that American Indians were capable of building such works. The Indian population seemed too small, their technology inadequate and their culture and nature did not appear suited to the organisation and labour required to perform such a mammoth undertaking. This belief was strengthened by the fact that when the Indians themselves were questioned, they invariably denied any involvement in their construction or function and either knew nothing of their origins or referred vaguely to a previous race (Silverberg, 1970, p. 11).

Various theories were proposed about the origins of the "mound builders". Among the suggested candidates were the Vikings, Toltecs, Phoenicians, Greeks, Romans, Persians, Hindus, Welsh and Danes (Birmingham & Eisenberg, 2000, p. 17). A popular and persistent theory was that the mounds were built by the lost tribes of Israel or that there was at least a Hebrew connection (Charles River Editors, 2015)

John Heckewelder

Considerable interest was generated when a missionary for the Moravian church, John Heckewelder, who had lived amongst the Delaware Indians from 1772, recounted some of their oral traditions. I have been unable to find Heckewelder's original publications but in an article printed in the Science newspaper, 16 May 1890, Cyrus Thomas quoted Heckewelder as follows:

> The Lenni Lenape (according to the tradition handed down to them by their ancestors) resided many hundred years ago in a very distant country in the western part of the American continent. For some reason which I do not find accounted for, they determined on migrating to the eastward, and accordingly set out together in a body. After a very long journey and many nights' encampment by the way, they at length arrived on the *Namaesi-Sipu*, where they fell in with the Mengwe, who had likewise emigrated from a distant country and had struck upon this river somewhat higher up. Their object was the same with that of the Delawares: they were proceeding on to the eastward until they should find a country that pleased them. The spies which the

Lenape had sent forward for the purpose of reconnoitring, had, long before their arrival, discovered that the country east of the Mississippi was inhabited by a very powerful nation, who had many large towns built on the great rivers flowing through their land. These people (as I was told) called themselves Talligew or Tallegewi ... Many wonderful things are told of this famous people. They are said to have been remarkably tall and stout; and there is a tradition that there were giants among them, people of much larger size than the tallest of the Lenape. It is related that they had built to themselves regular fortifications or intrenchments, from which they would sally out, but were generally repulsed. I have seen many of the fortifications said to have been built by them, two of which in particular were remarkable. One of them was near the mouth of the River Huron, which empties itself into the Lake St Clair on the north side of that lake, at a distance of about twenty miles north-east of Detroit. This spot of ground was, in the year 1776, owned and occupied by a Mr Tucker. The other works, properly intrenchments, being walls or banks of earth regularly thrown up, with a deep ditch on the outside, were on the Huron River, east of Sandusky, about six or eight miles from Lake Erie. Outside of the gateway of each of these intrenchments, which lay within a mile of each other, were a number of large flat mounds, in which, the Indian pilot said, were buried hundreds of the slain Tallegwi whom I shall hereafter, with Col. Gibson, call Allegewi. Of the intrenchments, Mr Abraham Steiner, who was with me at the time when I saw them, gave a very accurate description, which was published in Philadelphia in 1789 or 1790, in some periodical work the name of which I cannot at present remember.

When the Lenape arrived on the banks of the Mississippi, they sent a message to the Alligewi to request permission to settle themselves in their neighbourhood. This was refused them, but they obtained leave to pass through the country and seek a settlement further to the eastward. They accordingly began to cross the Namaesi-Sipu, when the Allegewi, seeing that their numbers were so very great, and in fact they consisted of many thousands, made a furious attack upon those who had crossed, threatening them all with destruction if they dared to persist in coming over to their side of the river. Fired at the treachery of these people and the great loss of men they had sustained, and, besides, not being prepared for a conflict, the Lenape consulted on what was to be done, whether to retreat in the best manner they could, or to try their strength and let the enemy see that they were not cowards, but men, and too high-minded to suffer themselves to be driven off before they had made a trial of their strength and were convinced that the enemy was too powerful for them. The Mengwe, who had hitherto been satisfied with being spectators from a distance, offered to join them on condition that after conquering the country they should be entitled to share it with them. Their proposal was accepted, and the resolution was taken by the two nations to conquer or die.

Having thus united their forces, the Lenape and Mengwe declare war against the Alligewi, and great battles were fought, in which many warriors fell on both sides. The enemy fortified their large towns and erected fortifications, especially on large rivers or near lakes, where they were successfully attacked and sometimes stormed by the allies. An engagement took place in which hundreds fell, who were afterwards buried in holes, or laid together in heaps and covered over with earth. No quarter was given, so that the Allegewi at last, finding that their destruction was inevitable if they persisted in their obstinancy, abandoned the country to the conquerers, and fled down the Mississippi River, from whence they never returned.

The war which was carried on with this nation lasted many years, during which the Lenape lost a great number of their warriors, while the Mengwe would always hang back in the rear, leaving them to face the enemy.

The purpose of sharing this account is not to suggest that nineteenth century belief in a lost race was universal. Even in the early days there were those who argued that the mounds were built by the native Indian tribes that were still living in North America. However, it is important to recognise that there were many notable Americans who espoused a belief in lost race on the North American continent. For example, in the auspicious year of 1830, the United States President, Andrew Jackson, gave his second annual message to Congress. In this speech Jackson tried to justify the possible extinction of the American Indian peoples with the following analogy:

> To follow to the tomb the last of his race and to tread on the graves of extinct nations excite melancholy reflections. But true philanthropy reconciles the mind to these vicissitudes as it does to the extinction of one generation to make room for another. In the monuments and fortresses of an unknown people spread over the extensive regions of the West, we behold the memorials of a once powerful race, which was exterminated or has disappeared to make room for the existing savage tribes (Marder, 2005, p. 100).

This was the society into which Joseph Smith Jnr. was born and raised. Given this, is it reasonable to assume that he was aware of stories of a lost race? We should also note that his home was located only a few miles from the Hill Cumorah in an area that contains physical evidences of ancient civilisation. When the angel Moroni appeared to Joseph Smith and told him of a book deposited nearby that gave an account of the former inhabitants of "this continent, and the source from whence they sprang", it may have made perfect sense to the young boy (The Testimony of the Prophet Joseph Smith).

Four years after Moroni's first visit, Joseph Smith received the plates on which the Book of Mormon was engraved. During the work of translation, he became well acquainted with the civilisations described in the Book of Mormon. His mother wrote:

> During our evening conversations, Joseph would occasionally give us some of the most amusing recitals that could be imagined. He would describe the ancient inhabitants of this continent, their dress, mode of traveling, and -the animals upon which they rode; their cities, their buildings, with every particular; their mode of warfare; and also their religious worship. This he would do with as much ease, seemingly, as if he had spent his whole life among them (Smith, 1853, p. 85).

In 1834, Joseph led a group of Latter-Day Saints from Kirtland, Ohio to Clay County, Missouri in a march that came to be known as Zion's Camp. The route traversed the entire width of territory that archaeologists now recognise as lands of the Hopewell Culture from east to west.

During the journey Joseph wrote a letter to his wife, Emma, which included the following:

> The whole of our journey, in the midst of so large a company of social, honest and sincere men, wandering over the plains of the Nephites, recounting occasionally the history of the Book of Mormon, roving over the mounds of that once beloved people of the Lord, picking up their skulls and their bones, as a proof of its divine authenticity, and gazing upon a country the fertility, the splendour, and the goodness so indescribable, all serves to pass away time unnoticed.

It appears that Joseph Smith Jnr., the prophet of the restoration and translator of the Book of Mormon, believed, or knew, that the mounds and bones that covered that part of North America belonged to the peoples described in the Book of Mormon. This was demonstrated in an incident that occurred during Zion's Camp. Diary accounts were recorded by six early members who were eye witnesses to the event, namely, Wilford Woodruff, Heber C. Kimball, George A. Smith, Levi Hancock, Moses Martin, and Reuben McBride.

In the course of their travels the saints had "visited several mounds which had been thrown up by the ancient inhabitants of this country—Nephites, Lamanites, etc." (Cannon, 1995). On Monday 2 June, 1834 the saints crossed the Illinois River. They camped on the river bank until the following day. In the morning a group of them walked up to the top of a high mound located near the river. The mound was high enough to provide a view over the tree tops and across the prairies on either side of the river to the distant horizon.

At the top of the mound the group found stones "which presented the appearance of three altars having been erected one above the other, according to the ancient order; and the remains of bones were strewn over the surface of the ground" (Cannon, 1995).

The group obtained a shovel and a hoe and dug about a foot into the mound before discovering a human skeleton with a stone arrow point lodged between the ribs. One of the members retained the arrow point. Joseph Smith's response to this incident was recorded as follows:

> The contemplation of the scenery around us produced peculiar sensations in our bosoms; and subsequently the visions of the

past being opened to my understanding by the Spirit of the Almighty, I discovered that the person whose skeleton was before us was a white Lamanite, a large, thick-set man, and a man of God.

His name was Zelph. He was a warrior and chieftain under the great prophet Onandagus, who was known from the Hill Cumorah, or eastern sea to the Rocky Mountains. The curse was taken from Zelph, or, at least, in part—one of his thigh bones was broken by a stone flung from a sling, while in battle, years before his death. He was killed in battle by the arrow found among his ribs, during the last great struggle of the Lamanites and Nephites. (Cannon, 1995).

Source of Illustration: Charles MacKay, The Mormons (1851)

Four years later, another expedition known as Kirtland's Camp, also made a connection between the northern United States and the Book of Mormon. The history of the Camp, later published in the Millennial Star, Vol. 16, p. 296, states: "The camp passed through Huntsville, in Randolph County, which has been appointed as one of the stakes of Zion, and is the ancient site of the City of Manti". Samuel D. Tyler, a member of the Camp, wrote in his journal "the ancient site of the City of Manti, which is spoken of in the Book of Mormon and this is appointed one of the Stakes of Zion, and it is in Randolph County, Missouri, three miles west of the county seat" (Joseph Fielding Smith, 1956, Vol. 3, p. 239).

When white settlers first arrived in the Mississippi Valley it contained more than 100,000 pre-Columbian mounds. The ancient North Americans built in timber and in earth and inhabited the most fertile and productive lands. Unfortunately, those lands were equally attractive to modern settlers. As a result, over the last two centuries much has been lost to the ravages of time and the spread of modern civilisation. Despite this, we must ask why such a rich and intriguing past has fallen so far from the consciousness of many people. Nineteenth century Americans were far more aware of this past than we are today. It is true that during the intervening years much physical evidence has been lost but it is also true that much research has been completed and the store of archaeological knowledge has greatly expanded.

The same question applies doubly to Latter-Day Saints. As noted in the Preface, during the current generation much effort has been dedicated to the study of Central and South America as a possible location for the Book of Mormon. Given the facts outlined above, including the statements of Joseph Smith and the location of the hiding place of the gold plates, has sufficient effort been put into the study of North America?

Rod Meldrum has argued that the text of the Book of Mormon itself provides additional reasons for considering North America. As you read the following quotes please consider whether they relate to a Central or South American nation or are more properly ascribed to the United States.

WAR OF INDEPENDENCE

1 Nephi 13:14 And it came to pass that I beheld many multitudes of the Gentiles upon the land of promise; and I beheld the wrath of God, that it was upon the seed of my brethren; and they were scattered before the Gentiles and were smitten.

16 And it came to pass that I, Nephi, beheld that the Gentiles who had gone forth out of captivity did humble themselves before the Lord; and the power of the Lord was with them.

17 And I beheld that their mother Gentiles were gathered together upon the waters, and upon the land also, to battle against them.

18 And I beheld that the power of God was with them, and also that the wrath of God was upon all those that were gathered together against them to battle.

19 And I, Nephi, beheld that the Gentiles that had gone out of captivity were delivered by the power of God out of the hands of all other nations.

USA BECOMES A WORLD POWER

1 Nephi 13:30 Nevertheless, thou beholdest that the Gentiles who have gone forth out of captivity, and have been lifted up by the power of God above all other nations, upon the face of the land which is choice above all other lands, which is the land that the Lord God hath covenanted with thy father that his seed should have for the land of their inheritance; wherefore, thou seest that the Lord God will not suffer that the Gentiles will utterly destroy the mixture of thy seed, which are among thy brethren.

1 Nephi 22: 7 And it meaneth that the time cometh that after all the house of Israel have been scattered and confounded, that the Lord God will raise up a mighty nation among the Gentiles, yea, even upon the face of this land; and by them shall our seed be scattered.

2 Nephi 10: 12 And I will fortify this land against all other nations.

A LAND WHICH IS CHOICE ABOVE ALL OTHER LANDS

2 Nephi 1: 5 But, said he, notwithstanding our afflictions, we have obtained a land of promise, a land which is choice above all other lands; a land which the Lord God hath covenanted with me should be a land for the inheritance of my seed. Yea, the Lord hath covenanted this land unto me, and to my children forever, and also all those who should be led out of other countries by the hand of the Lord.

A LAND OF LIBERTY

2 Nephi 1: 7 Wherefore, this land is consecrated unto him whom he shall bring. And if it so be that they shall serve him according to the commandments which he hath given, it shall be a land of liberty unto them; wherefore, they shall never be brought down into captivity; if so, it shall be because of iniquity;

for if iniquity shall abound cursed shall be the land for their sakes, but unto the righteous it shall be blessed forever.

LAMANITES SCATTERED

1 Nephi 13: 14 And it came to pass that I beheld many multitudes of the Gentiles upon the land of promise; and I beheld the wrath of God, that it was upon the seed of my brethren; and they were scattered before the Gentiles and were smitten.

NEW JERUSALEM

3 Nephi 20: 22 And behold, this people will I establish in this land, unto the fulfilling of the covenant which I made with your father Jacob; and it shall be a New Jerusalem. And the powers of heaven shall be in the midst of this people; yea, even I will be in the midst of you.

Ether 13: 3 And that it was the place of the New Jerusalem, which should come down out of heaven, and the holy sanctuary of the Lord.

4 Behold, Ether saw the days of Christ, and he spake concerning a New Jerusalem upon this land.

5 And he spake also concerning the house of Israel, and the Jerusalem from whence Lehi should come—after it should be destroyed it should be built up again, a holy city unto the Lord; wherefore, it could not be a new Jerusalem for it had been in a time of old; but it should be built up again, and become a holy city of the Lord; and it should be built unto the house of Israel—

6 And that a New Jerusalem should be built up upon this land, unto the remnant of the seed of Joseph, for which things there has been a type.

We should consider whether the preceding verses relate to Mexico or another Central American country or whether they refer to the United States of America. If they do more aptly describe the latter then they give support to the study of North America as a possible location for the Book of Mormon.

We should also consider the angel Moroni's reference to "this continent" when he visited Joseph Smith and spoke about the gold plates. Do those words, uttered in New York state, indicate that the Book of Mormon is a history of North America?

Amberli Nelson proffered yet another reason for considering North America (Nelson, 2012). In the Book of Mormon we read:

> 2 Nephi 5:10 And we did observe to keep the judgements, and the statutes, and the commandments of the Lord in all things, according to the Law of Moses.

Keeping the Law of Moses in all things, as described in the Old Testament, requires the use of various animals and plants including ox, bull, ram, lamb, goat, dove, turtledove or pigeon, flour, olive oil, and wine. Davies argued that if the Lord required the Nephites to obey the Law of Moses, he would have ensured that the requisite plants and animals were available in their new land. She pointed out that all the aforementioned plants and animals are native to North America and are absent from Central America.

A major proponent of the Central American theory, John L. Sorenson (1985), aware of the complete absence of such animals and plants from Central or South America suggested other animals and plants that were endemic to that region could be used as substitutes for those named in the Law. So, for example, instead of lamb or ram, he listed "camelidae, paca, or *agouti*" (p. 299). Camelidae is a group of animals related to the camel which includes Alpaca and Llama, Paca is a large rodent and agouti is also a rodent. All of these animals are classified as unclean under the Law of Moses. The suggestion that unclean animals would be used as substitutes to keep the Law of Moses in all things, or that a large relative of the rat could be used as a substitute for a lamb to symbolise the Saviour in sacred ordinances must be questioned. If it was agreed that such substitutions would not be in keeping with the Law of Moses then this would also point us towards North America.

When a paradigm is established it generally takes a lot of effort to disrupt that paradigm and move in a different direction. I have outlined some of the arguments that disrupted my paradigm in relation to Book of Mormon geography and encouraged further thought in relation to that subject. My hope is that in the chapters that follow you will find information that resolves your questions and helps you to discover a new, more reassuring paradigm.

2

The Archaic Period

NORTH AMERICAN CULTURES AND TRADITIONS

According to Silverberg (1970), "a culture is a specific social group with a distinct way of life" (p. 171). American archaeologists use the term "culture" but they also commonly use the word "tradition". Tradition may indicate a culture that persists over a longer time period or larger geographical area or it may indicate a series of cultures that share similarities. The term tradition "is used especially to designate specific New World cultures such as the ... Woodland Tradition ... The attributes, styles, traits, or technologies develop continuously, thus forming an easily accounted-for series of advancements" (Archaeology Wordsmith, n.d.).

It should be noted that cultures and traditions are based on things which can and have been examined, such as pottery styles, not on the people who created those things and have long since passed away. Changes in traditions may result from the evolution of technologies used by the same peoples or they can mark a change in the ethnic groups that populate an area. It should be emphasised that depicting traditions or cultures as a continuum in time can be misleading. For example, Fortier (2001) when discussing the American Bottom argued that the vertical nature of chronological charts "is unfortunate, for it implies connectivity" (p. 192). The true situation was very different with that region demonstrating "a hodgepodge of cultural traditions, many of which are horizontally intrusive, unrelated, and short-lived" (Fortier, 2001, p. 192). While the various traditions in the American Bottom from Early Woodland to Late Woodland are chronological, they are not connected and therefore the cultural changes witnessed in that area are not the result of an evolutionary process (Fortier, 2001, p. 192).

It should not therefore be automatically assumed that cultures and traditions represent changes in a single ethnic population. It must be understood that lands can be abandoned by one ethnic group and later populated by another, as we shall shortly see.

The time period that encompasses the Book of Mormon begins in what is termed by archaeologists as the "late archaic period". The archaic period ends around 1,000 B.C. and because of the nature of archaic societies and the many years that have elapsed, the knowledge of what happened at that time is incomplete. As such, I will not try to provide a comprehensive overview of archaic North America but, rather, I will draw the readers' attention to two significant cultures, i.e., Poverty Point Culture and Old Copper Culture.

POVERTY POINT CULTURE (2200-600 B.C.)

The Poverty Point Culture is named after a site at Poverty Point, Louisiana, in the Mississippi River valley. The site includes six semi-elliptical earthen ridges with an outer diameter of 1.14 km and a large flat plaza bordered by the ridges. About 35,000 cubic metres of soil were used in the construction of the earthworks. This was the largest and most significant community associated with the culture and yet its purpose is not yet understood. Kidder (2008) explained that the people of Poverty Point and related sites were consuming "vast" amounts of goods, often brought in from great distances but they had very little to exchange in return, perhaps only "a handful of jasper owl beads" (p. 30). Kidder asked: "What was Poverty Point exchanging as a currency for the goods it acquired? What could Poverty people export?" (p. 30) This question will be addressed later.

The culture extended well beyond Poverty Point throughout the lower Mississippi Valley and surrounding Gulf coast. According to Kidder (2008) there were high population densities throughout the Late Archaic period which demonstrated "extensive settlement diversity, widespread long-distance trade and exchange, seemingly complex behavioural patterns manifest in mound architecture, burial patterns, and grave inclusions, and considerable artefact diversity" (p. 24).

Archaeologists do not know what brought about the demise of the Poverty Point Culture. In 2006 Kidder proposed climate change as a possible cause, however, in 2008, he recanted that idea (Kidder, 2008). While no one seems prepared to offer an alternative solution, Anderson (2008) provided his opinion on the possible role of warfare. He stated that though he did not know for sure he felt it unlikely that warfare was the cause of their demise. His reasoning for this was that they had not found evidence of the sacking of Poverty Point or any of the other major sites or of the inhabitants being massacred (p. 289).

While there is no evidence of war at Poverty Point or the other major centres there is, nevertheless, evidence of war. Anderson (2008) conceded that "appreciable evidence for weapons trauma is observed" where "well-preserved human remains" have been discovered by researchers such as Smith (1996, cited in Anderson, 2008) and therefore there probably was some type of "regular or recurring conflict" (Anderson, 2008, p. 289).

A number of different time periods have been suggested for the Poverty Point Culture. Wikipedia proposed 2200 B.C.–700 B.C.; Oxford Reference 1700 B.C.–700 B.C.; Neuman and Hawkins (1993) 2000 B.C.–600 B.C.; while Kidder (2008) and Gibson (2008) gave 1,000 B.C. as the year the culture ended.

The end of the Poverty Point Culture was "sudden and marked" (Kidder, 2008, p. 31) and its disappearance was followed by a "five-to-seven century gap in history" (Gibson, 2008, p. 33). Gibson (2008) reported that "it was as if a giant hand swept away humanity for half a millennium" (p. 38). Further, Kidder (2008) noted that "while there was a gap in occupation in much of the interior riverine Southeast, a similar pattern had been noted for parts of northeastern United States" (p. 23).

In Summary, the Poverty Point Culture existed from 2,200 B.C. or later until 600 B.C. or earlier. It was a thriving culture with high population densities and consumed vast quantities of goods imported over long distances. We don't know what it produced or why it mysteriously came to an end but when it did it left an empty landscape, devoid of population.

OLD COPPER CULTURE (MINING 2400-1200 B.C.)

The terms "Old Copper Culture" and "Old Copper Complex" refer to "items made by early inhabitants of the Great Lakes region during a period that spans several thousand years and covers several thousand square miles" (Cullen, 2006, para. 1). Who created those items has been the subject of much debate and controversy over the past two centuries but there are facts upon which most people agree.

Copper has been a valuable metal mined and used in various cultures for millennia. Most copper is found mixed with other minerals and must be physically separated. This process was very difficult and laborious prior to the invention of modern machinery. However, there are deposits of copper, known as native copper, which do not need to undergo the separation process. Native copper is a natural mineral which is not combined with other materials and was of particular value to ancient peoples. The largest and perhaps purest deposits of native copper in the world are found around the Great Lakes of North America.

Huge quantities of copper were mined in the Great Lakes region in both modern and ancient times. During 100 years of modern mining about 10 million tons of copper were removed and "beneath each and every modern copper mine was an ancient pit mine" (Hoffman, 2014). There are more than 5,000 ancient copper mines in the Isle Royale, Keweenaw Peninsula area of Michigan. While there has been no definitive study to calculate the quantity of copper mined anciently, estimates put the figure as high as 50 million tons.

Ancient miners extracted the copper from the rock in which it was embedded by a process known as spalling. Spalling involved building fires under the material, heating it to high temperatures and then dousing it with cold water. The sudden change in temperature caused the surrounding rock to crack away from the copper. Extracting large quantities of copper required a large workforce over an extended period. Mertz (2004) mused that to undertake such large-scale mining over such a big area using such a "crude and slow process" over hundreds of years the work would have to be "carried on by a vast number of people" (p. 44).

After centuries of mining the end of the ancient industry was both sudden and inexplicable. There have been multiple modern finds of large pieces of copper, weighing as much as six tons, still sitting in ancient mine pits supported by ancient timbers where they had been abandoned millennia ago. It was as if a large workforce was going about its labours one day and simply did not return to work the next. What adds to the mystery is that there is no evidence of such vast quantities of copper ever being used in the Great Lakes region.

Where the copper went is the subject of debate and controversy. Graham Hancock has been active in researching and promoting one possible solution to this question while the majority of archaeologists reject Hancock's assertions as lacking in solid evidence. Hancock's theory will be touched upon later.

As noted above, copper has been used by many cultures. In the Great Lakes region pieces, known as float copper, were picked up on the surface and used by local tribes over thousands of years, but when did the industrial scale mining occur? Mertz (2004) stated that two carbon 14 tests at one pit had provided results "of 3800 +- 300 and 3000 +- 300 or dating this pit at 1800 B.C. to 1000 B.C. with a possible variation of 300 years earlier or later" (p. 39). He went on to state:

> With comparative accuracy provided by these carbon 14 datings, indications point to a possibility that working of copper mines on Michigan's Upper Peninsula could well have taken place as early as 2000 B.C.-a period of time fantastically irreconcilable with archaeological determinations of a pre-carbon 14 era (p. 40).

In a later study, Wakefield (2015) stated that the mining continued from 2400 B.C. till 1200 B.C.

The Copper Culture was involved in large-scale mining during the same time period that the Poverty Point Culture was thriving at the other end of the Mississippi. The Copper Culture produced vast quantities of copper for an unknown market and like the Poverty Point Culture, came to a mysterious end.

THE JAREDITES (2200-600 B.C.)

According to the Book of Mormon, to the north of the Nephite homelands was a land the Nephites called "Desolation". Helaman stated that It was so named because of the destruction of the people who had previously occupied that land.

> Helaman 3:3 And it came to pass in the forty and sixth, yea, there was much contention and many dissensions; in the which there were an exceedingly great many who departed out of the land of Zarahemla, and went forth unto the land northward to inherit the land.

> 4 And they did travel to an exceedingly great distance, insomuch that they came to large bodies of water and many rivers.

> 5 Yea, and even they did spread forth into all parts of the land, into whatever parts it had not been rendered desolate and without timber, because of the many inhabitants who had before inherited the land.

6 And now no part of the land was desolate, save it were for timber; but because of the greatness of the destruction of the people who had before inhabited the land it was called desolate.

7 And there being but little timber upon the face of the land, nevertheless the people who went forth became exceedingly expert in the working of cement; therefore they did build houses of cement, in the which they did dwell.

8 And it came to pass that they did multiply and spread, and did go forth from the land southward to the land northward, and did spread insomuch that they began to cover the face of the whole earth, from the sea south to the sea north, from the sea west to the sea east.

9 And the people who were in the land northward did dwell in tents, and in houses of cement, and they did suffer whatsoever tree should spring up upon the face of the land that it should grow up, that in time they might have timber to build their houses, yea, their cities, and their temples, and their synagogues, and their sanctuaries, and all manner of their buildings.

10 And it came to pass as timber was exceedingly scarce in the land northward, they did send forth much by the way of shipping.

It must be stressed that while the land described above lacked trees it was not a desert. As stated, "no part of the land was desolate, save it were for timber" (Helaman 3:6). The land was capable of producing crops, feeding herds, supporting a large human population and growing trees. An earlier expedition to the land northward described more than just the vegetation.

> Mosiah 8:8 ... a land among many waters, ... a land which was covered with bones of men, and of beasts, and was also covered with ruins of buildings of every kind, having discovered a land which had been peopled with a people who were as numerous as the hosts of Israel.

This treeless land was a stark contrast to the lands immediately to their south.

> Alma 22:31 ... the land on the southward was called Bountiful, it being the wilderness which is filled with all manner of wild

animals of every kind, a part of which had come from the land northward for food.

The expedition mentioned above also found a record left by the former inhabitants of the land. An abridgement of that record was included in the Book of Mormon under the name *Book of Ether*. Unfortunately, the abridgement contains only 34 verses that relate to activities in and characteristics of the land. Such a brief account was never going to provide detailed information about culture and location, however, it does reveal that the people lived beyond the area known as Desolation. One of their kings lived in the land of Moron "near the land which is called Desolation by the Nephites" (Ether 7:6). Another king "departed out of the land with his family, and travelled many days, and came over and passed by the hill of Shim, and came over by the place where the Nephites were destroyed, and from thence eastward, and came to a place which was called Ablom, by the seashore ..." (Ether 9:3).

The place where the Nephites were destroyed refers to the Hill Cumorah, in upstate New York (Mormon 8:2). The Hill Cumorah was also the location of the final battle in which the last of the people, known as Jaredites, were destroyed. The book also states that it took four years to gather all the people together (to Cumorah) for the final conflict. This would indicate a population that was geographically widely dispersed.

The land of Desolation may however have been the heartland. The Book of Ether states that the land southward was preserved as "a wilderness, to get game" while "the whole face of the land northward was covered with inhabitants" (Ether 10:21).

When reading the Book of Ether, it is possible to see connections between the geographical features mentioned in the text and the Great Lakes region. Key connections are:
- the story ends at Cumorah, in upstate New York. We know where that is;
- on route to that destination reference is made to sea south, sea north, sea west and sea east. When reading this, it is important to be aware of the distinction between "seas" and "oceans". The Bible refers to the Red sea, Sea of Galilee and Mediterranean Sea or Great Sea, none of which are oceans: They are all bodies of water that are surrounded by land. The Book of Mormon always uses different terms such as "many

waters" or "great waters" when referring to oceans. When we consider that the people travelled northward to a region where there were at least four seas it could be referring to the four Great Lakes;

- it refers to a narrow neck of land. This may be the Niagara area or to the area between Lake Michigan and the former Great Black Swamp, west of Lake Erie;
- it talks of the place where the sea divides the land. The Great Lakes form a barrier stretching hundreds of miles that truly divides the land;
- it describes a land among many waters. The State of Michigan has over 11,000 lakes and no place is more than 10km from a lake.

Readers may also see the link with mining. Mining is not an activity that is often mentioned in scripture, yet the Book of Ether talks about large scale mining and processing of metals:

> Ether 10:23 And they did work in all manner of ore, and they did make gold, and silver, and iron, and brass, and all manner of metals; and they did dig it out of the earth; wherefore, they did cast up mighty heaps of earth to get ore, of gold, and of silver, and of iron, and of copper. And they did work all manner of fine work.

The connection between the Great Lakes region and ancient copper mining on a massive scale was discussed earlier. It should also be noted that gold, silver, iron and zinc (a component of brass) are all found in large quantities in Michigan. Gold has been found in over 100 places in Michigan. Silver is often found with copper. Three of the six largest iron deposits in the United States are found in Michigan. Zinc is also mined commercially in Michigan. While mineral deposits are found in other regions of the Americas the glacial history of this region, the variety of minerals found in such a relatively small area, and the magnitude of ancient mining, make it quite unique.

Any linkage between the absence of timber mentioned in the Book of Mormon and the Great Lakes region may be a little more difficult to establish. When Europeans first arrived in Michigan it was covered with dense stands of timber and by 1870 this became the basis of a booming timber industry. However, by 1910 the trees had almost vanished and the timber trade ceased. Is it possible that the ancient forests had experienced a similar fate? Could human activities over many centuries account for the lack of timber described in the Book of Mormon? As previously stated, timber was an essential component of the ancient mining process, in spalling, scaffolding and transporting copper in carts therefore it must have been available to the ancient miners. It was almost certainly an important resource in the construction of ancient buildings for a population which "covered the whole face of the land northward" (Ether 10 :21). Timbered areas may have also been cleared for farming activities. The fact that the Jaredites made a policy to preserve the more southern area would tend to indicate that the absence of trees that the Nephites discovered was a result of Jaredite activities.

Did the Nephites begin the reafforestation of the land of Desolation? As previously mentioned, when the Nephites began to settle in the land of Desolation they used alternate building products and "they did suffer whatsoever tree should spring up upon the face of the land that it should grow up" (Helaman 3:9). The two thousand years that followed was surely sufficient time for the forests of Michigan to re-establish themselves.

The ancient large-scale mining in the Great Lakes region fits within the timeframe of the Book of Ether. The book states that the Jaredites came to their promised land after the confusion of languages at the Tower of Babel, or around 2200 B.C. The last of the Jaredites died sometime after 600 B.C. As noted above, the copper mines of Michigan date from around 1800 B.C. to 1000 B.C. give or take 300 years.

The Book of Ether may explain why ancient copper mining around the Great Lakes suddenly ceased. It devotes 167 verses to the reign of kings and their wars. At some point in time continual warfare rendered normal life impossible as more and more of the population was pressed into military service and other activities ceased.

> Ether 14:21 And so great and lasting had been the war, and so long had been the scene of bloodshed and carnage, that the whole face of the land was covered with the bodies of the dead.
>
> 22 And so swift and speedy was the war that there was none left to bury the dead, but they did march forth from the shedding of blood to the shedding of blood, leaving the bodies of both men, women, and children strewed upon the face of the land, to become a prey to the worms of the flesh.

I would ask readers to consider whether the Book of Ether may also resolve an unanswered question referred to earlier, that is, the disappearance of the Poverty Point Culture. Archaeologists do not link the Copper Culture and Poverty Point Cultures, however, they existed during the same time period at opposite ends of the same river system, they were both involved in activities that have not been fully explained and they both ended mysteriously at about the same time.

Until archaeologists can fill in the blanks, I think we are entitled to speculate. Was the settlement at Poverty Point, in the lower Mississippi valley, linked in any way to the copper trade or other activities around the Great Lakes and northern Mississippi valley? Was the ongoing Jaredite warfare responsible for "the appreciable evidence for weapons trauma that is observed in at least some parts of the region" described by Anderson (2008, p. 289)? Likewise, is the reason that no evidence has been found "to suggest that Poverty Point or any other major sites and centres of the terminal Archaic were sacked and their inhabitants massacred" (Anderson, 2008, p. 289) because the southern population was gathered to a final conflict far to the north, at Cumorah?

3

Early Woodland
(600–100 B.C.)

GIBSON (2008) DESCRIBED THE LOWER Mississippi valley and surrounding coastal areas at the end of the Archaic Period as "if a giant hand swept away humanity for half a millennium" (p. 38). Into this uninhabited land came the Early Woodland Culture.

The Early Woodland Culture has not received a great deal of attention from archaeologists. Williams, (1963, as cited in Kidder, 2008, p. 24) called it the epitome of a "good gray culture". The interior southeast and much of the Northeast is "ignored, treated in a few sentences, or discussed only in relation to the important processes of interest" (Kidder, 2008, p. 24). Despite this paucity of research, archaeologists have provided information that is of value. Archaeologists describe the Woodland Period as one of "continuous development in stone and bone tools, leather crafting, textile manufacture, cultivation, and shelter construction" (Woodland Period, n.d.). This period included the widespread use of pottery and the increasing use of horticulture which led to the development of villages and cities.

Kidder (2008) examined the Early Woodland Culture and provided a number of important insights. He compared it with the earlier Poverty Point Culture and found, in general terms, that the Early Woodland Culture was simpler, it had simpler architecture, burial practices and artefacts. It also had smaller populations, fewer types of settlements and less long-distance trade (p.24).

A number of differences were also noted within the Early Woodland Culture. Kidder (2008) found that on the eastern side of the Mississippi River there was evidence of an "increase in Early Woodland settlement in upland regions". [In contrast,] "site densities west of the Mississippi River or in the uplands adjacent to western tributaries (e.g., the Ouachita) do not appear to increase at all during the Early Woodland. In fact, in many of these regions Early Woodland components are nearly non-existent" (p.29).

Thirdly, Jenkins, Dye and Walthall (1986, as cited in Kidder, 2008) described the diffusion of one of the key markers of Early Woodland Culture through the lower Mississippi valley. They wrote, "one of the critical markers of Early Woodland, namely pottery, diffused or was carried out of the lower Southeast and moved in a westerly and northerly direction along the coast and into and along the major river valleys (p. 29)."

THE LEHITES (589 B.C.—385 A.D.)

The Book of Mormon records that two extended families arrived by ship in the south of the promised land in around 589 B.C. There is nothing in the text to suggest that when they landed there were any other inhabitants. In fact, Nephi wrote:

> 2 Nephi 1:8 And behold, it is wisdom that this land should be kept as yet from the knowledge of other nations; for behold, many nations would overrun the land, that there would be no place for an inheritance.

These two families lived together by the coast for a relatively short period of time before the party divided into two groups, namely, the Nephites and the Lamanites. One group, the Nephites, travelled north for "many days" before settling in a place they named the land of Nephi. The Nephites remained in the land of Nephi until sometime between 279 B.C. and 124 B.C. During that time there was warfare between the Nephites and the Lamanites. It may have been around 225 B.C. when a portion of the Nephites left their lands to the Lamanites and travelled further north to the land of Zarahemla.

The text provides some details about Lamanite settlements. When people travelled from the land of Zarahemla to the land of Nephi they "went up to the land" (Alma 17:8). As the land of Nephi was south of Zarahemla we can assume that "up" referred to elevation, i.e. land of Nephi was not in a river valley, as was the city of Zarahemla, but in an upland area further away from the river.

Some Lamanites lived in villages whereas others lived a more hunter gatherer lifestyle.

Alma 22:28 Now, the more idle part of the Lamanites lived in the wilderness, and dwelt in tents; and they were spread through the wilderness on the west, in the land of Nephi; yea, and also in the west of the land of Zarahemla, in the borders by the sea-shore, and on the west in the land of Nephi, in the place of their father's first inheritance, and thus bordering along by the seashore.

The preceding few paragraphs provide important congruence between the archaeological record and the Book of Mormon.

Evidence	Archaeological Record	Book of Mormon
Timing	Kidder (2008) and Gibson (2008) placed the beginning of the Woodland Period at 500 B.C. When we consider the tiny numbers that arrived with Lehi, a date of 500 B.C. is acceptable. Neuman and Hawkins (1993) proposed 600 B.C.	Around 589 B.C.
Other inhabitants	According to Kidder (2008) and Gibson (2008) the lower Mississippi valley and southern coast was uninhabited for several centuries before 589 B.C. This is in stark contrast to other parts of the Americas, including Central America, where populations existed uninterrupted for many centuries before, during and after that date	Land kept from the knowledge of other nations.
Direction of movement	Settlement proceeded from the south to the north.	Landed on the southern coast and moved north.
Settlement patterns	Initial settlement was concentrated in upland areas. Those who lived in the west pursued a more primitive lifestyle.	Land of Nephi "up". Idle Lamanites lived in the west.
Continuous development	The development of the society in terms of sophistication and population size was what would be expected from a small, new community as time progressed.	Two extended families increased in numbers and sophistication.

ADENA CULTURE (600 B.C.-150 A.D.)

The first chapter of this book referred to the discovery of tens of thousands of ancient structures in the Mississippi valley. The Adena Culture, along with the Hopewell Culture, built the vast majority of those edifices. Their efforts "far exceeded those of the other cultures, and it is the mounds and earthworks of these two groups that have made such a distinctive and lasting contribution to the prehistoric cultural landscape of the region" (Woodward & McDonald, 2001, p. 15).

There are differing views as to when precisely the Adena Culture commenced. Charles River Editors (2015) wrote "...the so-called "Mound Builder Tradition" emerged from the upper Ohio River Valley about 700 B.C. known as the Adena (Chapter 2). Romain (2015) reported "some archaeologists believe that Adena extends as far back as 800 B.C. Others posited a beginning date of about 500 B.C. (p. 21)." Woodward and McDonald (2001) stated that the "Adena might not have appeared until several centuries after 1000 B.C. (p. 24)." These different sources provide a starting date somewhere between 500 and 800 B.C.

Different dates have also been proposed for the end of the Adena Culture. However, it did not come to an end in the same way that other cultures, such as Poverty Point, did. Charles River Editors (2015) explained that although the Adena Culture was geographically large and "well-established" it would somewhat pale beneath what was to come. By around 150 B.C. the Adena would be dominated by the extraordinary Hopewell Culture. The Hopewell took Adena practices and made them even more elaborate. Over time, "the Adena slowly faded into the cultural landscape—their accomplishments usurped and absorbed by this next great society" (Chapter 2).

The relationship between the Adena and the Hopewell is still a matter of discussion, however, some characteristics are generally agreed. Most archaeologists accept that the Adena were the first to establish themselves in the northern Mississippi valley. The Hopewell were later arrivals who moved into the region previously controlled by the Adena. After their arrival there was "close contact and even intermarriage between Adena and Hopewell" (Silverberg, 1970, p. 199). While the Hopewell population was smaller than the Adena it came to be the dominant culture and as previously noted, the Adena culture faded into the background.

The territory of the Adena "extended from southeastern Indiana to southwestern Pennsylvania, and from north central Ohio to central Kentucky" (Woodward & McDonald, 2001, p. 7). It is generally accepted that the Hopewell culture originated in Illinois before spreading into Adena habitats in Ohio and other states.

A note of caution. These views are based on the balance of available information and may or may not change as new information comes to hand. For example, when radiocarbon dating was first used to compare occupation dates of both Adena and Hopewell sites in the Ohio Valley with an Illinois site, the Hopewell sites tested were older than the Adena sites and the Illinois site was the oldest. Spaulding's response to this was "radiocarbon dates will be ignored in this discussion, and the earlier view of the basic position of Adena will be retained" (Spaulding, 1952, p. 260).

The cultural characteristics that defined the Adena included:

- The use of pottery;
- The development of agriculture. The Adena propagated a number of native plants including goosefoot, knotweed, little barley, maygrass, sumpweed, sunflower, and squash (Woodward & McDonald, 2001, p. 26);
- The construction of earthen mounds. Some of these mounds were for burials, others contained no human remains and may have served as territorial markers;

- The construction of sacred enclosures (usually an earthen circle) and palisaded villages.

The skeletal remains found in Adena mounds were physically different from other cultures and included women over six feet in height and men approaching seven feet (Silverberg, 1970, p. 195). These physical differences along with cultural differences led Spaulding (1952) to conclude that the Adena culture was not a diffusion of new practices into an existing population but was the result of the migration of new people into the Ohio Valley (p. 264). Where these people came from is the subject of speculation.

BURIAL MOUNDS

The construction of Adena burial mounds was generally a four-step process. As described by Woodward and McDonald (2001), it would begin by preparing "a special surface or floor" (p. 29). The deceased would then be prepared for burial. This may involve de-fleshing the body and applying red ochre to the bones. The deceased would then be placed in a chamber, either at ground level or below. Simple grave goods would normally be placed with the body. A mound would then be built "over and around the grave" (p. 29). It should be noted that this process was reserved for a special few. According to Spaulding (1952) "the mound building procedure outlined is plainly confined to a special class, presumably of great prestige" (p. 262).

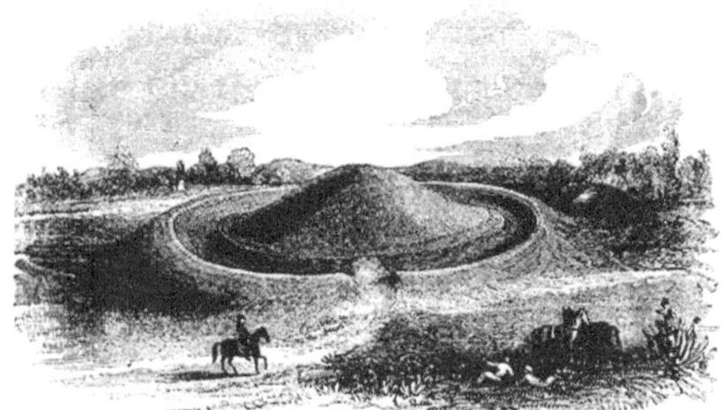

Source of Illustration: *Ancient Monuments of the Mississippi Valley,* Squier & Davis (1988)

Burial mounds, otherwise called tumuli, have been used by many cultures throughout many parts of the world including Europe and the Middle East. The Israelites built at least 19 such structures near Jerusalem. Whether Judean kings were interred in these mounds, or tumuli, is speculation as no inscriptions have been excavated. However, when one tumulus was hastily excavated to make way for a construction project, two LMLK seal impressions and two handles with Concentric Circle incisions were found. LMLK seals are ancient Hebrew seals stamped on the handles of large storage jars dating from reign of King Hezekiah (circa 700 B.C.). This suggests that the tumulus belonged to either King Hezekiah (Barkay, 2003, p. 68) or his son Manasseh (Grena, 2004, p. 326; see, also, Tumulus, n.d., Wikipedia)

THE MULEKITES (587-120 B.C.)

Sometime between 279 B.C. and 124 B.C., perhaps around 225 B.C., a portion of the Nephites travelled north and discovered "the people of Zarahemla". According to the Book of Mormon they were descendants of "Mulek, and those who came with him into the wilderness" (Mosiah 25:2). Mulek was a son of the Jewish King, Zedekiah. The Old Testament tells us that when the Babylonians captured Zedekiah, they put all his sons to death, but the Book of Mormon account states that one son, Mulek, and his companions:

> Omni 1:15 ... came out from Jerusalem at the time that Zedekiah, king of Judah, was carried away captive into Babylon. (587 or 586 B.C.)
>
> 16 And they journeyed in the wilderness, and were brought by the hand of the Lord across the great waters ...

Unlike the Nephites, who landed in the south of the promised land and travelled northwards, Mulek, and his companions were brought into "the land North (Helaman 6:10)". They first landed in the land of Desolation, the Jaredite heartland, and travelled south (Alma 22:30). During the time between arriving in the promised land and being discovered by the Nephites things had not been all smooth sailing. "... They had had many wars and serious contentions, and had fallen by the sword from time to time ..." and, without records, their language had become corrupted and they had lost any religious beliefs (Omni 1:17).

By the time that the Nephites discovered the people of Zarahemla "they had become exceedingly numerous" (Omni 1:17). They were more numerous than the Nephites and yet despite this "the people of Zarahemla, and of Mosiah [Nephites], did unite together; and Mosiah was appointed to be their king" (Omni 1:19). Nearly a century later, at about 120 B.C., the Nephites and the people of Zarahemla "gathered together in two bodies" (Mosiah 25:4) which indicates that though they had united under one king, they had up until that point in time maintained their separate cultural identities.

Unfortunately, that is all we know about the Mulekites. Who travelled with Mulek or how they travelled from Israel to America is not revealed. In regard to the time period of perhaps 350 years from their arrival in America till the people of Zarahemla were discovered by the Nephites, all we know is that they had many wars. We do not even know who those wars were between, whether they were civil wars or if they involved other peoples. We don't know if the people of Zarahemla were the only Mulekites then in America or whether any groups had separated and moved into other areas. Further, we can only speculate as to why the Mulekites accepted a Nephite as their king even though they clearly outnumbered the Nephites and their number may have included direct descendants of King Zedekiah. After the gathering in 120 B.C. they were referred to as Nephites in the Book of Mormon.

SPECULATION

Before comparing what we do know about the Mulekites with the archaeological record, let us indulge briefly in speculation about how Mulek and his group crossed the great waters. While the author Amaleki acknowledged the Lord's hand in the journey, it may not have been the same type of involvement that we saw in Lehi's journey. Mulek was a son of an unrighteous king who would not listen to the Lord's prophets. His group took with them no sacred records and after arriving in the new land their descendants "denied the being of their creator" (Omni 1:17). So perhaps the journey depended more upon the knowledge of those involved than upon divine revelation.

In Chapter 2 we talked about the Copper Culture around the great lakes and noted that no one knows what happened to the massive amounts of copper that were mined there anciently. The theory promoted by Graham Hancock, which is not popular with archaeologists, is that the ore was taken across the Atlantic to satisfy the huge demand for copper in Europe during the Bronze Age. If that was the case, the sailing route was a secret well kept by one of the ancient seafaring nations, such as the Phoenicians. The reason I raise this speculation is that the Mulekites did not land in the south, where the Lord had led the Lehites, but in the north, in the land of Desolation, where the Jaredites had been involved in large scale mining. Perhaps this landing site was more than just a coincidence. Perhaps Mulek, son of a Jewish King, took advantage of an existing trade route to escape the fate that befell his siblings.

THE ARCHAEOLOGICAL RECORD

Though the Book of Mormon contains scant information about the Mulekites it is interesting to compare what we do know with the archaeological record:

- Timing—the Mulekites arrived sometime after 587 B.C. This date fits within the range of dates proposed by leading archaeologists for the commencement of the Adena culture;
- Intrusion—the Mulekites intruded into their new territory in the manner that Spaulding (1952) believes the Adena came into the northern Ohio valley;
- Order of arrival—the Mulekites arrived in their new territory first and were later joined by and co-existed with the Nephites.
- The Nephites, though fewer in number, dominated the relationship. This matches the order of arrival and interaction of the Adena and Hopewell as described by Charles River Editors and others;
- Timing of Nephite arrival—the time of the arrival of the Nephites into Mulekite lands correlates with the arrival of the Hopewell in Adena lands; and
- Construction of burial mounds—the Book of Mormon does not describe Mulekite burials but it is possible that the burial of important persons in mounds was a practise of Kings in Jerusalem at that period of time and was brought to the Americas by Mulek, son of a Jewish king.

4

Middle Woodland

HOPEWELL CULTURE (200 B.C.-400 A.D.)

The Hopewell culture was the "main event" of North American pre-history. It was responsible for the majority of the earthen mounds that generated so much debate in the nineteenth Century. These included burial mounds, gigantic geometric structures with astronomical alignments, temple mounds and massive fortifications. The Hopewell acquired exotic goods through an extensive trade network and their influence came to dominate much of the Mississippi Valley. As a result, much archaeological work has gone into studying the evidences left behind by this extraordinary culture.

Source of Illustration: *Ancient Monuments
of the Mississippi Valley,* Squier & Davis (1988)

We should not forget that when on Zion's camp Joseph Smith wrote he was "wandering over the plains of the Nephites" (letter to Emma Smith) he was in Hopewell territory. When he was "roving over the mounds" (letter to Emma Smith) he was treading upon Hopewell or Adena mounds and when he was picking up skulls and bones, if they were from within the timeframe of the Book of Mormon, they were Hopewell or Adena skulls and bones.

Because of the large amounts of information available from archaeological sources and from the Book of Mormon, this section will be different to earlier chapters. Instead of waiting until the end of the section to draw parallels between archaeology and scripture, discussion will follow each topic.

TIMING

Archaeologists date the Hopewell culture in the Illinois River and Ohio River valleys between 200 B.C. and 400 A.D. (Birmingham & Eisenberg, 2000, p.84). The earlier date relates to the Illinois river valley while 100 B.C. is the "date often cited as the beginning of Hopewell in Southern Ohio" (Carskadden & Morton, 1997, p. 371). While the Hopewell spread into other areas, Illinois and Ohio were the focal points of the civilisation.

In 1841 Joseph Smith received a revelation which instructed the saints to "build up a city unto my name upon the land opposite the city of Nauvoo, and let the name of Zarahemla be named upon it" (D&C 125:3). The site of this city was in the centre of the Havana Hopewell territory. If this was the site of ancient Zarahemla, the Book of Mormon account would agree with the archaeological record in three respects:

1. According to the Book of Mormon the Nephites arrived in the land of Zarahemla sometime between 279 B.C. and 124 B.C., perhaps around 225 B.C. This timing matches the date proposed by Birmingham and Eisenberg (2000);

2. The Nephites "began to scatter abroad upon the face of the earth, yea, on the north and on the south, on the east and on the west (Mosiah 27:6)" probably between 100 and 92 B.C. This scattering matches the date proposed by Carskadden and Morton (1997); and

3. The order in which the settlement occurred also agrees.

The end of the Hopewell culture will be discussed in later sections but for now let us add:

4. The final destruction of the Nephite people occurred in 385 A.D., which aligns with the date proposed by Birmingham and Eisenberg (2000).

AGRICULTURE

It has been accepted since their discovery, that a society which could construct the works completed by the Hopewell, was an agricultural society. In more recent times more information about the Hopewell diet has been uncovered. Riley, Edging, and Rossen (1990) list a number of native plants that the Hopewell consumed. They included goosefoot, marshelder, sunflower, knotweed, maygrass, little barley, squash, gourds, and tobacco. This was a similar diet to the Adena however the Hopewell were involved in growing these crops more intensively and unlike the Adena also grew some maize (Riley et al., 1990).

The Book of Mormon described their agricultural pursuits as follows:

Mosiah 9:9 And we began to till the ground, yea, even with all manner of seeds, with seeds of corn, and of wheat, and of barley,

and with neas, and with sheum, and with seeds of all manner
of fruits; and we did begin to multiply and prosper in the land.

The Book of Mormon account refers to corn and barley, which were
part of the Hopewell diet. It also refers to neas and sheum which are
words that we are not familiar with. Perhaps these were names given to
some of the native plants that the Nephites were unfamiliar with in the
old world.

It is also notable that the Nephite currency system was based on grain
and in particular barley noting that "a senum of silver was equal to a
senine of gold, and either for a measure of barley, and also for a measure
of every kind of grain" (Alma 11:7). It could be inferred that barley was a
staple crop which would accord with the abundance of little barley in the
Hopewell territories.

BURIAL

At least three different burial methods were used by the Hopewell.
First, Birmingham and Eisenberg (2000) noted that since 1987, when
the Burial Sites Preservation Law was enacted, several mass graves that
are suspected to be from the Middle Woodland period have been discovered away from any mound sites along the Mississippi River (p. 95). At
Fort Ancient, Randall (1908) reported "human bones in vast quantities—
'bushels of them'—were found here a few inches below the surface soil"
(pp. 101-102).

Second, some Hopewell dead were interred in crypts. According to
Brown (1979) unlike the more complex burial mounds, Hopewell crypts
are basically large boxes that were built to hold the dead and their accom-
panying grave goods. The burial process was very simple, with the bodies
or bones of the deceased being placed in the crypt with little fanfare. The
crypts appear to be communal space with dead of the same "size, age, and
sex profiles" as the general population sharing space randomly (p. 211-2).
They were located away from the village on high spots of ground where
they were buried and covered to protect them from the possibility of scav-
engers getting in. This was a much easier and simpler burial process and
would appear to be more common than mound burials. (Brown, 1979,
p. 211-2)

Third, like the Adena, the Hopewell also constructed burial mounds. Mound burials commenced with the construction of a building, known as a charnel house, where the human remains were deposited. The bodies were physically prepared and it is possible that ceremonies were performed within the charnel house. Later, the charnel house was burned and a mound built over it. Only a small percentage of the population were interred in burial mounds. The "form and elaborateness" of their burial was a "declaration" of their station and status in life. (Brown, 1979, pp. 211-2).

The Book of Mormon provides no information on individual burials. It does record that after one large battle "their dead bodies were heaped up upon the face of the earth, and they were covered with a shallow covering" (Alma 16:11). After another battle we read "the bodies of many thousands are laid low in the earth, while the bodies of many thousands are moldering in heaps upon the face of the earth" (Alma 28:11). This record may be in line with the mass graves that have been discovered. The simple crypt burials would not be out of place with peoples of Hebrew origin. The mound burials may have originated with the Adena, as previously discussed, and been adopted by the Hopewell for those of high status, particularly at times of pride as recorded in the Book of Mormon.

BUILDINGS

The Hopewell lived in one of the largest hardwood forests on the planet and so their homes and community buildings were constructed of timber. Public buildings were generally rectangular but sometimes circular or irregular shaped and multi-chambered but always made of timber. Walls may have been timber or a combination of earth topped with wooden palisades. Any other structures would also have been timber. "As long as a site, or a part of a site, was in use, wooden architecture probably dominated the landscape" (Woodward & McDonald, 2001, p. 53).

The same was true of domestic buildings. Woodward and McDonald (2001) reported that at the Stubbs site a lot of new information was uncovered about the diversity of house like structures built during the second century A.D. Many different floor plans were found with postholes which again confirmed the importance of timber to the Hopewell residents (Woodward & McDonald, 2001, p. 249).

Unfortunately, Hopewell wooden structures did not have the longevity of stone buildings and have long ago rotted and crumbled. Even when they were in existence they may not have been as splendid as the stone constructions we see in Central and South America. However, the important aspect of Hopewell buildings is not their grandness but how they align with the text of the Book of Mormon which unambiguously explains that timber was the building material used by the Nephites.

The Book of Mormon records that the first Nephite leader, Nephi, taught his people "to build buildings, and to work in all manner of wood" (2 Nephi 5:15).

Two hundred years later Jarom reported, "and we multiplied exceedingly, and spread upon the face of the land, and became exceedingly rich in gold, and in silver, and in precious things, and in fine workmanship of wood, in buildings, and in machinery ..." (Jarom 1:8).

Wicked King Noah "built many elegant and spacious buildings; and he ornamented them with fine work of wood, and of all manner of precious things, of gold, and of silver, and of iron, and of brass, and of ziff, and of copper; And he also built him a spacious palace, and a throne in the midst thereof, all of which was of fine wood and was ornamented with gold and silver and with precious things. And he also caused that his workmen should work all manner of fine work within the walls of the temple, of fine wood, and of copper, and of brass (Mosiah 11:8-10).

When the people spread into northern areas that were devoid of trees, they were careful to "suffer whatsoever tree should spring up upon the face of the land that it should grow up, that in time they might have timber to build their houses, yea, their cities, and their temples, and their synagogues, and their sanctuaries, and all manner of their buildings" (Helaman 3:9). To alleviate the scarcity of timber "they did send forth much by way of shipping" (Helaman 3:10).

EARTHQUAKES

Throughout this Chapter direct comparisons have been drawn between Middle Woodland Cultures and the text of the Book of Mormon. The discussion on earthquakes will vary from this pattern. This section will begin with descriptions of earthquakes that have occurred in historical times because this is the most detailed and extensive information that we have in relation to earthquakes in North America's heartland. It should be noted, however, that evidence has been found that supports similar earthquakes occurring in the prehistoric period. For example, Stover and Coffman (1989) have written that the uplift of the Tiptonville dome was caused by earthquakes that occurred between 200 and 2,000 years ago, or possibly within the Middle Woodland period. So, while historic earthquakes will be discussed comparisons will be drawn with events that are described in the Book of Mormon.

Between the 16th December 1811 and 23rd January 1812 three massive earthquakes occurred with their epicentres in Missouri. In terms of magnitude they were all ranked in the top seven to affect mainland USA in modern recorded history, but in terms of geographical area affected, these were the largest. Fortunately, the area was sparsely populated at the time.

"The region most seriously affected covered an area of 78,000 - 129,000 square kilometres, extending from Cairo, Illinois, to Memphis, Tennessee, and from Crowley's Ridge in northeastern Arkansas to Chickasaw Bluffs, Tennessee" (Stover & Coffman, 1989, p. 68). "Chimneys were toppled and log cabins were thrown down as far distant as Cincinnati, Ohio, St. Louis, Missouri, and in many places in Kentucky, Missouri, and Tennessee" (Stover & Coffman, 1989, p. 68). Eliza Bryan was living in New Madrid on the night of the first quake. She reported:

> ... about two o'clock, a.m., we were visited by a violent shock of an earthquake, accompanied by a very awful noise resembling loud but distant thunder, but more hoarse and vibrating, which was followed in a few minutes by the complete saturation of the atmosphere, with sulphurous vapor, causing total darkness. The screams of the affrighted inhabitants running to and fro, not knowing where to go, or what to do—the cries of the fowls and beasts of every species—the cracking of trees falling, and the roaring of the Mississippi— the current of which was retrograde for a few minutes, owing as is supposed, to an irruption in its bed—formed a scene truly horrible. (1811-12 New Madrid earthquakes, n.d.)

The effects of the quakes included soil liquification, "general ground warping, ejections, fissuring, severe landslides, and caving of stream banks" (Stover & Coffman, 1989, p. 262). "Coal and sand were ejected from fissures in the swamp land adjacent to the St. Francis River, and the water level is reported to have risen there by 8 to 9 meters" (Stover & Coffman, 1989, p. 68). "Thousands of fissures ripped open fields, and geysers burst from the earth, spewing sand, water, mud and coal high into the air" (Rusch, E., 2011). "The skies turned dark during the earthquakes, so dark that lighted lamps didn't help. The air smelled bad, and it was hard to breathe. It is speculated that it was smog containing dust particles caused by the eruption of warm water into cold air" (Strange Happenings during the Earthquakes, n.d.).

"The earthquakes caused the ground to rise and fall - bending the trees until their branches intertwined and opening deep cracks in the ground" (Stover & Coffman, 1989, p. 66). "Deep seated landslides occurred along the steeper bluffs and hillsides; large areas of land were uplifted permanently; and still larger areas sank and were covered with water that erupted through fissures or craterlets" (Stover & Coffman, 1989, p. 68).

Source of Illustration:
The Great Earthquake at New Madrid,
a 19th-century woodcut from
Devens, Our First Century (1877)

Over a period of five months the affected area experienced more than 2,000 aftershocks. The people living in the area discovered that the crevices which were caused by the quakes generally ran in a north south direction and so devised a strategy to safeguard themselves from being swallowed up in the earth. When the earth started to quake, the residents would cut down trees in an east west direction and hold on to the trees until the shaking ceased in the hope that this would save them from the crevices. Unfortunately, this method was not always successful. There were "missing people" who were most likely swallowed up by the earth. Some earthquake fissures were as long as five miles" (Strange Happenings during the Earthquakes, n.d.).

"Huge waves on the Mississippi River overwhelmed many boats and washed others high onto the shore. High banks caved and collapsed into the river; sand bars and points of islands gave way; whole islands disappeared" (Stover & Coffman, 1989, p. 68). Also, uplift created temporary waterfalls waves that propagated upstream. Subsidence of 1.5 to 6 metres formed Reelfoot Lake in what is now Lake County, Tennessee (1811-12

New Madrid earthquakes, n.d.) and Lake St Francis in eastern Arkansas (Stover & Coffman, 1989, p. 68).

As noted earlier, this was not the first major earthquake to affect the Midwest. Evidence has been found of prehistoric earthquakes that have made permanent changes to the landscape. In southwest Kentucky, southeast Missouri and northwest Tennessee a region known as The Lake County has been lifted by as much as 10 metres above the surrounding countryside. The affected area is 50 km long and 23 km wide. The area of highest uplift, known as Tiptonville dome, is 11 km long and 14 km wide. On the eastern side of the dome is the 3 m high Reelfoot scarp. It is believed that the vertical movement occurred during earthquakes between 200 and 2,000 years ago. (Stover & Coffman, 1989, p. 68).

The region's history of earthquakes, as significant as it is, is not the only geological activity that sets it apart. The northern areas of land around the Great Lakes are well known in the scientific community for some unusual geological features. During the last ice age, the Great Lakes region was covered by an immense ice sheet that was one kilometre thick. The advance and subsequent melting of this ice sheet changed the shape of the Great Lakes through several iterations, creating different lake stages and different water flows. Even today the region is undergoing a process known as isostatic rebound. That is, when the enormous weight of the ice was removed, the land began to rise and fall on either side of hinge points. This movement is evidenced by ancient beaches that stand above the current water line in some areas and plunge below the water line in others.

The Algoma stage of the Great Lakes ended about 2,000 years ago. Between the Algoma Stage and the present-day lakes was the Sault stage. The Sault stage "is represented by beaches on the north shore of Lake Superior that have been preserved by crustal uplift still in progress. Elsewhere around the Superior basin the Sault beaches are submerged" (Wright & Frey, 2015, p. 80).

The combination of geological conditions described above is very unusual. There, in the Hopewell heartland, is the site of the largest earthquake in United States' history in terms of area of land affected. Alongside that, in the northern parts of their territory, is an area of isostatic rebound that has caused changes in land levels thrusting some lands below the water and lifting other lands up.

The Book of Mormon records a major disaster that occurred in 33 A.D.

> 3 Nephi 8:9 And the city of Moroni did sink into the depths of the sea, and the inhabitants thereof were drowned.
>
> 10 And the earth was carried up upon the city of Moronihah, that in the place of the city there became a great mountain
>
> 12 But behold, there was a more great and terrible destruction in the land northward; for behold, the whole face of the land was changed, because of the tempest and the whirlwinds, and the thunderings and the lightnings, and the exceedingly great quaking of the whole earth;
>
> 13 And the highways were broken up, and the level roads were spoiled, and many smooth places became rough.
>
> 14 And many great and notable cities were sunk, and many were burned, and many were shaken till the buildings thereof had fallen to the earth, and the inhabitants thereof were slain, and the places were left desolate.
>
> 15 And there were some cities which remained; but the damage thereof was exceedingly great, and there were many in them who were slain.
>
> 17 And thus the face of the whole earth became deformed, because of the tempests, and the thunderings, and the lightnings, and the quaking of the earth.
>
> 18 And behold, the rocks were rent in twain; they were broken up upon the face of the whole earth, insomuch that they were found in broken fragments, and in seams and in cracks, upon all the face of the land.

The destruction described here was not the result of a minor event. Areas of land sank below the sea (don't forget the meaning of sea), other areas were lifted up and, in the land northward, the whole face of the earth was deformed. Also, note that the destruction was not localised. It was a widespread event affecting all the Nephite lands.

The Book of Mormon also describes dreadful noise and fissures opening up and swallowing people.

> 3 Nephi 10:9 And it came to pass that thus did the three days pass away. And it was in the morning, and the darkness dispersed

from off the face of the land, and the earth did cease to tremble, and the rocks did cease to rend, and the dreadful groanings did cease, and all the tumultuous noises did pass away.

10 And the earth did cleave together again, that it stood ...

14 ... all these deaths and destructions by ... the opening of the earth to receive them ...

There are many parallels between the 1811 and 1812 earthquakes and the destruction described in the Book of Mormon. They were all major earthquakes in terms of severity and the size of the area affected. They all damaged and destroyed buildings and roads. They all caused large vertical movements in the land, as happened at Tiptonville and other areas. They all created horrible noises and caused fissures to open up and swallow people. We are also entitled to speculate whether the forces of a major earthquake in combination with isostatic rebound caused the city of Moroni and other cities to be sunk beneath the waters.

These are not the only parallels we find in connection with the destruction.

3 Nephi 8:16 And there were some who were carried away in the whirlwind; and whither they went no man knoweth, save they know they were carried away.

Whirlwinds that can carry people away may today be referred to as tornados. The Mid-West of the United States is known as tornado alley because of its propensity for tornados. Tornadoes are most common in the middle latitudes and do not generally occur in equatorial regions.

The most unusual phenomenon recorded in the Book of Mormon in connection with the destruction was a period of total darkness upon the face of the earth.

20 And it came to pass that there was thick darkness upon all the face of the land, insomuch that the inhabitants thereof who had not fallen could feel the vapor of darkness;

21 And there could be no light, because of the darkness, neither candles, neither torches; neither could there be fire kindled with their fine and exceedingly dry wood, so that there could not be any light at all;

22 And there was not any light seen, neither fire, nor glimmer, neither the sun, nor the moon, nor the stars, for so great were the mists of darkness which were upon the face of the land.

Darkness is not usually associated with earthquakes. The darkness may cause some speculation as to whether the shaking was caused by volcanic activity and the darkness by ash or a pyroclastic flow. However, volcanic ash would not cause the total darkness and inability to light even candles and pyroclastic flows are deadly to humans.

The 1811-1812 earthquakes, on the other hand, produced the world's largest sand boil. "Thousands of fissures ripped open fields, and geysers burst from the earth, spewing sand, water, mud and coal high into the air" (Rusch, E., 2011). There were multiple reports of darkness so complete that lamps were useless.

In summary, the destruction described in the Book of Mormon can only be described as extraordinary. It included an earthquake of great size and intensity which caused damage and destruction to buildings and roads, uplift and sinking of lands, and the opening and closing of fissures which swallowed people. It was associated with the sinking of lands below seas and with whirlwinds or tornados.

These extraordinary events raise a number of important questions. First, how many other areas in the Americas have the geological and climatological features to enable such happenings to occur? Second, how many other areas have provided archaeological and/or historical evidence of earthquakes of great magnitude along with all the characteristics described in the Book of Mormon? Third, how many other areas have produced evidence of activity during the relevant time period?

EXPANSION

The expansion of the Hopewell culture into Wisconsin and southeastern Minnesota, Missouri, and Michigan did not come through the adoption of Hopewell practices by a resident population. Kingsley (1981) wrote:

> It is the position of this paper that the bulk of the data at hand supports the hypothesis that Hopewell in west Michigan was indeed the result of a direct movement of people from Illinois (p. 155).

He went on to write that there was very little evidence for a previous tradition in that area, that is, the Hopewell moved into previously unoccupied territory.

Kingsley expanded the scope of his statement by citing another researcher: "Griffin has posited population expansion out of the Illinois centre into southern Wisconsin and southeastern Minnesota, Missouri, and Michigan. The intrusion of Illinois Havana peoples seems to have occurred by at least AD 100 (Kingsley, 1981, p. 163)."

These propositions match the Book of Mormon account that the Nephites "began to scatter abroad upon the face of the earth, yea, on the north and on the south, on the east and on the west" (Mosiah 27:6) probably between 100 and 92 B.C. It also supports the possibility that the land had been "kept as yet from the knowledge of other nations" (2 Nephi 1:8).

FORTIFICATIONS

In the first half of the 19th century, before modern civilisation had caused the loss of so much of the prehistoric, Squier and Davis set out on a mission to record the ancient monuments of the Mississippi Valley. With a clearer and more complete view than is possible today, they found hundreds of structures of earth and rock that, in their opinions, were of a military nature. For example: "Nothing can be more plain, than that most of the remains in northern Ohio, particularly those on the Cuyaoga river, are military works" (Squier & Davis, 1988, p. 41). They went further to state, "there seems to have existed a System of Defences extending from the sources of the Alleghany and Susquehanna in New York, diagonally across the country, through central and northern Ohio, to the Wabash" (Squier & Davis, 1988, p. 44). In their opinion, these defences were constructed in response to a protracted contest in which the builders of the mounds were "constantly exposed to attack" (Squier & Davis, 1988, p. 44).

Other archaeologists have also described lines of defence built by the Hopewell.

> Three great works on the Great Miami, one at its mouth, one at Colerain, and one at Hamilton, with subsidiary defensive works extending six miles along the river at Hamilton; several advanced

works to north and west of Hamilton, on streams flowing into the Great Miami; and other similar defences farther up the river at Dayton and Piqua, all put in communication with each other by signal mounds erected at conspicuous points, constitute together a connected line of defences along the Miami river; Fort Ancient on the Little Miami stands as a citadel in the rear of the centre of this line. A mound at Norwood, back of Cincinnati, commands a view through a depression of the hills at Redbank eastwardly to a mound in the valley of the Little Miami; northwardly through the valley of the Millcreek and the depression in the land thence to Hamilton, with the works at Hamilton; and by a series of mounds (two of which in Cincinnati and its suburbs have been removed) westwardly to the river, a distance of over one hundred miles, could transmit by signals an alarm from the little work north of Worthington through the entire length of the valley to the works at Portsmouth (Randall, 1908, p. 60).

The Book of Mormon records that the Nephites had fortified their cities to some degree as early as 399 B.C. However, in around 72 B.C., in response to an increasing threat, they greatly enhanced their defences. Their military leader "caused them to erect fortifications that they might secure their armies and their people ..." (Alma 50:10) and continued "fortifying the line between the land of Zarahemla and the land of Nephi..." (Alma 50:11).

FORTIFIED TOWNS AND CITIES

The Hopewell and Adena built fortifications around their towns and cities. They took advantage of the topography wherever possible and in other areas built walls for protection. These walls were built by excavating an external trench and using the soil to construct an earthen or earthen and stone wall. This wall could then be topped with a timber palisade.

For example, Woodward and McDonald (2001) wrote that "Peter Village was a palisaded enclosure with an exterior ditch bordering the wall ..." (p. 26). "The embankment was bordered on the outside by a ditch about 15 feet wide and four to eight feet deep ..." (p. 111).

Silverberg (1970) wrote: "Here were their towns, often surrounded by earth embankments..." (p. 151).

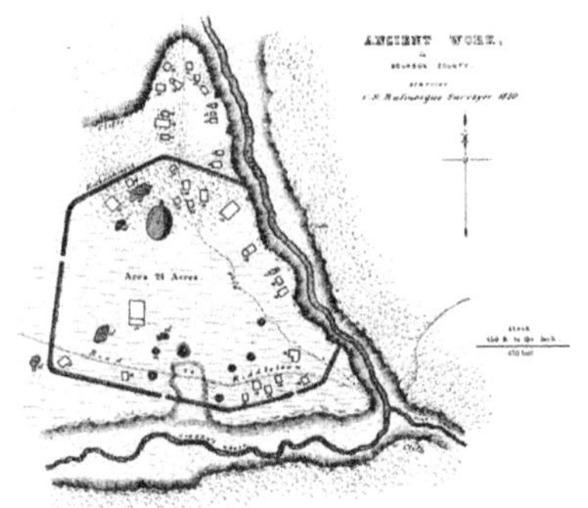

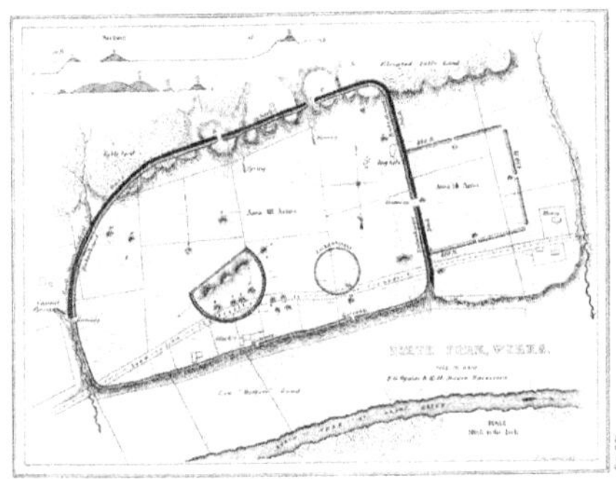

Source of Illustrations:
Ancient Monuments of the Mississippi Valley,
Squier & Davis (1988)

The Book of Mormon describes how the Nephites fortified their cities.

> Alma 50:1 ... they should commence in digging up heaps of earth round about all their cities ...
>
> 2 And upon the top of these ridges of earth he caused that there should be timbers, yea, works of timbers built up to the height of a man, round about the cities.
>
> 3 And he caused that upon those works of timbers there should be a frame of pickets built upon the timbers round about; and they were strong and high.
>
> 4 And he caused towers to be erected that overlooked those works of pickets, and he caused places of security to be built

upon those towers, that the stones and the arrows of the Lamanites could not hurt them.

5 And they were prepared that they could cast stones from the
top thereof, according to their pleasure and their strength, and
slay him who should attempt to approach near the walls of the
city.

PLACES OF RETREAT

In addition to fortifications around towns the Hopewell built places
of retreat atop high points on the edge of the Appalachian Plateau or in
other isolated spots where the natural terrain provided a natural defence.
These forts were never permanently occupied but were designed to be
a refuge in times of danger. Farming activities took place in the fertile
soils of the valleys. Permanent houses and day to day living all happened
in the valleys. Prufer (1997) stated that "during raids or whatever, kids,
women, etc. may have been shunted up to the top of the structure" (p.
320). He speculated that this is the reason that the fort sites contain
so little evidence of ongoing occupation. "... the consensus is that these
structures belong in the Hopewell complex of Middle Woodland times,
dated conservatively in Ohio from about 100 B.C. to AD 350" (Prufer,
1997, p. 314).

The forts were large, covering many acres and contained their own
water supply. They took advantage of their high positions and often
steep approaches and were further protected by walls of earth which were
sometimes reinforced with stone and may include a wooden palisade to
further raise the walls.

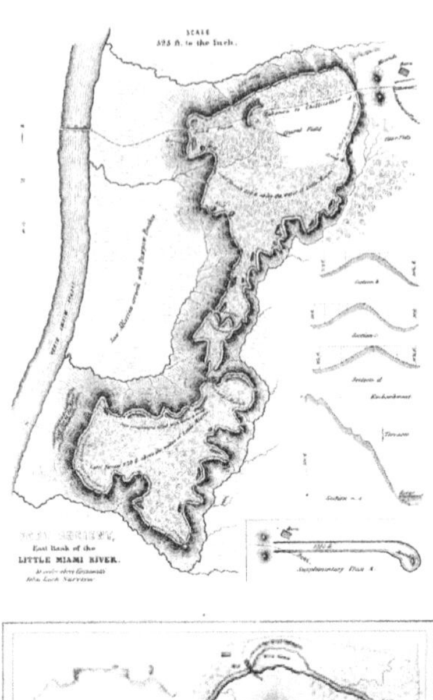

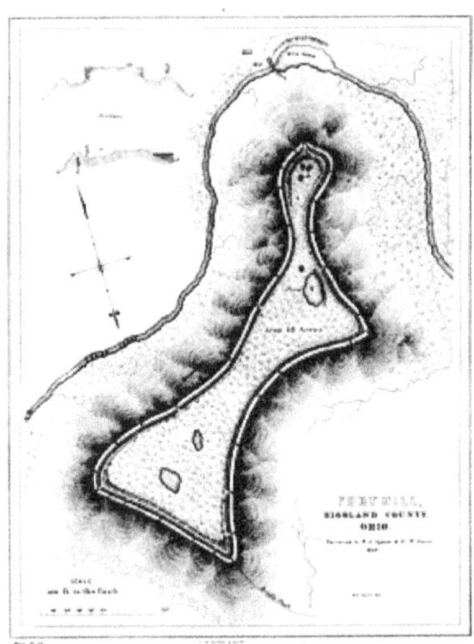

Source of Illustrations:
Ancient Monuments of the Mississippi Valley,
Squier & Davis (1988)

Fischer (1965 as cited in Prufer, 1997) investigated Fort Miami in Hamilton County. He found that the walls of that fort had been made of "earth, stone, and wood" (p. 316). He also noted that it appeared that at least one of the walls had at one time included a wooden palisade which was destroyed by fire. The ash from the fire was radio carbon dated at 270 A.D. plus or minus 130 years.

Riordan investigated a fort at the Pollock earthworks in Greene County. Like Fischer, Riordan found that the Pollock earthwork had also included a wooden palisade which had been destroyed by fire and subsequently replaced by an earthen embankment. Six radio carbon dates from that site gave dates ranging from 132 A.D. to 236 B.C. (cited in Prufer, 1997, p. 316).

The walls of these forts were not always continuous but some contained gaps at irregular intervals. It is believed that "each of these gaps was occupied with a blockhouse reaching out beyond the wall, forming a bastion from which defenders could enfilade the outside of the ramparts most effectually" (Randall, 1908, p. 96). "This exterior 'platform' might be used as a lookout or sentinel stand, and especially would this landing be of advantage when located on the edge of an inaccessible declivity" (Randall, 1908, p. 98).

Some forts had unusual entrances. There were no gates but openings that led to narrow maze-like passageways with several dead ends. This prevented those wishing to enter from doing so on masse and exposed them to the weapons of the defenders as they tried to find the correct path.

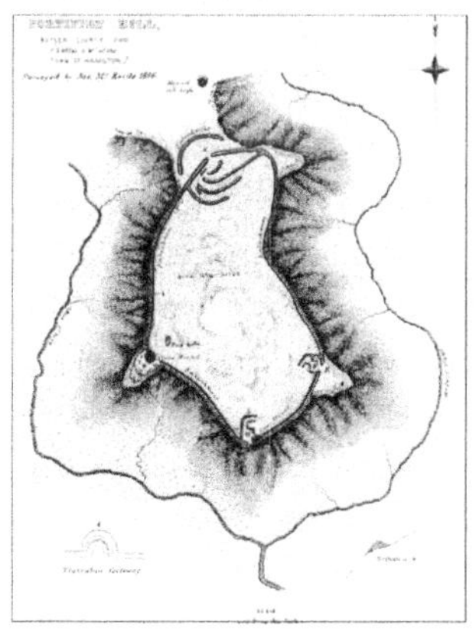

Source of Illustration:
Ancient Monuments of the Mississippi Valley,
Squier & Davis (1988)

The Book of Mormon matches the archaeological record in great detail. It describes how the Nephites erected:

> Alma 48:8 ... small forts, or places of resort; throwing up banks of earth round about to enclose his armies, and also building walls of stone to encircle them about, round about their cities and the borders of their lands; yea, all round about the land.

Every Nephite city was either fortified or a "fort of security" was built (Alma 49:3)

The book describes one attempt to storm such a fort.

> Alma 49:4 But behold, how great was their disappointment; for behold, the Nephites had dug up a ridge of earth round about them, which was so high that the Lamanites could not cast their stones and their arrows at them that they might take effect, neither could they come upon them save it was by their place of entrance.
>
> Alma 49: 18 Now behold, the Lamanites could not get into their forts of security by any other way save by the entrance, because of the highness of the bank which had been thrown up, and the depth of the ditch which had been dug round about, save it were by the entrance.
>
> 19 And thus were the Nephites prepared to destroy all such as should attempt to climb up to enter the fort by any other way, by casting over stones and arrows at them.
>
> 20 Thus they were prepared, yea, a body of their strongest men, with their swords and their slings, to smite down all who should attempt to come into their place of security by the place of entrance; and thus were they prepared to defend themselves against the Lamanites.
>
> 21 And it came to pass that the captains of the Lamanites brought up their armies before the place of entrance, and began to contend with the Nephites, to get into their place of security; but behold, they were driven back from time to time, insomuch that they were slain with an immense slaughter.
>
> 22 Now when they found that they could not obtain power over the Nephites by the pass, they began to dig down their banks of earth that they might obtain a pass to their armies, that they might have an equal chance to fight; but behold, in these

attempts they were swept off by the stones and arrows which were thrown at them; and instead of filling up their ditches by pulling down the banks of earth, they were filled up in a measure with their dead and wounded bodies.

Note the similarities between the Book of Mormon description for the construction of places of retreat and the method used by the Hopewell. Note how the Lamanites were exposed and driven off when trying to breach the walls of the structure as would be expected if attempting to storm a Hopewell wall with 'blockhouses'. Also, note the term 'place of entrance', the Lamanites unsuccessful attempts to enter through the 'place of entrance' and how it could describe the unusual maze-like entrances constructed at some Hopewell forts.

PEARLS

The Hopewell harvested fresh-water pearls from streams both within and without their home territories. The pearls were made into beads and sewn onto clothes or strung into chains. They were a sign of wealth and status and were sometimes used as grave goods. Around 15,000 pearls were found in the Seip-Pricer Mound covering a multiple burial. Thirty-five thousand were found in one place at the Turner Site in Hamilton County and more than 100,000 were found in the Hopewell group of mounds (Woodward & McDonald, 2001, p. 67).

Compare the archaeological findings with the text of the Book of Mormon

> 4 Nephi 1: 24 And now, in this two hundred and first year there began to be among them those who were lifted up in pride, such as the wearing of costly apparel, and all manner of fine pearls, and of the fine things of the world.

METALWORK

Metalwork was an important aspect of Hopewell culture though the sophistication of technology used is a matter of debate. Goods made of hammered and embossed copper and gold (Woodward & McDonald, 2001, p. 66), silver and meteoric iron have been recovered from Hopewell mounds. These items are beautifully crafted.

There is also some evidence that the Hopewell used other metals that have not survived until the present day. Squier and Davis (1988) questioned how they could produce such beautiful stone sculptures from such tough materials without using such materials. "It resists the best tempered blade and yields reluctantly to the finest grit-stones. Yet it is clear from the markings on certain specimens that it was cut by some kind of implement" (Squier & Davis, 1988, p. 286).

Another early researcher, Caleb Atwater, found evidence of the possible use of iron near two skeletons in a mound in Circleville. Beside the skeletons was a rusted metal plate which "seemed to resemble cast iron" along with "a handle made of elk's horn, at one end of which were traces of rust indicating the onetime presence of the iron blade of a sword or large knife" (Silverberg, 1970, p. 55).

More controversially, Marder (2005) believes that ancient furnaces have been found. He stated that 130 iron or copper-smelting furnaces have been discovered in Ohio, centred in Ross County and "a few in Virginia, Georgia, and Kentucky, and as far west as Arizona and New Mexico" (Marder, 2005, p. 27-28). Amateur archaeologist Arlington Mallery wrote that these furnaces resembled ancient Old World pit smelters (Mallory & Roberts-Harrison, 1979).

Such discoveries align with the Book of Mormon

> 2 Nephi 5:15 And I did teach my people to build buildings, and to work in all manner of wood, and of iron, and of copper, and of brass, and of steel, and of gold, and of silver, and of precious ores, which were in great abundance.

RELIGION

Archaeologists generally agree that religion was an important aspect of the Hopewell culture. "In many scholars' minds, there is little doubt that the Mound Builders were a pious people whose society was founded on spiritual beliefs and religious adherence on a massive scale" (Charles River Editors, 2015, Chapter 1).

There has been conjecture that the Hopewell brought their religion to the Adena people. Prufer (1964) speculated that the Hopewell religion was brought to Ohio from Illinois by "a privileged minority who in some way had come to dominate Ohio's Adena people".

It is important to stress that the religion was not the showy idol worship or polytheism of central America. Squier and Davis (1988) reported that in all their investigations they had not recovered or become aware of any relics "which were obviously designed as idols or objects of worship" (Squier & Davis, 1988, p. 243). They compared the highly sculptured and imposing relics of central America with the plain and simple structures of the Hopewell and stated that the Hopewell "have no sculptured facades of temples and palaces, invested with a symbolic meaning or commemorative of the exploits of chiefs or conquerors, nor have we ponderous statues of divinities and heroes" (Squier & Davis, 1988, p. 242).

The Book of Mormon records the story of a group of people from Jerusalem who travelled to the Americas under prophetic leadership. The Nephites observed the law of Moses until the old law was replaced by the gospel of Christ. They took this religion to the people of Zarahemla who by that time "denied the being of their creator" (Omni 1:17). The archaeological record reflects this transmission of religion from the Hopewell to the Adena and the simpleness of their structures, void of any evidence of idolatry, is what may be expected from a Judeo/Christian society.

RENAISSANCE

Sometime in the first half of the first century A.D., a dramatic change spread throughout the American Midwest. For the next 150 years an energised Hopewell culture brought renewal to all the inhabitants of the region. The archaeological evidence shows "fundamental changes occurred in technological, subsistence, community, ideological, and settlement systems in this area" (Fortier, 2001, p. 183).

The time from 50 A.D. to 200 A.D., referred to as the Holding phase, was a golden period for the Hopewell. Their culture spread throughout the Midwest and brought with it changes in technology and in settlement patterns. "Sweeping changes in both lithic and ceramic technology appear overnight, and smaller settlements are replaced by more extensive and nucleated horticulturally based villages" (Fortier, 2001, p. 183). Similar changes are observable throughout the Midwest at the same time.

Along with the changes in settlement patterns came changes in agricultural practices. "Food plants such as starchy seeds, wild bean, panicum type 61, as well as cucurbits, papaw, plum, persimmon, and small-seeded legumes increase dramatically ... Maize and tobacco appear for the first time (Fortier, 2001, p. 189).

Archaeologists know that the period from around A.D. 50 to A.D. 200 was a special period in the Midwest but they don't know what brought this about. The Book of Mormon may provide a useful insight.

> 3 Nephi 11:8 And it came to pass, as they understood they cast their eyes up again towards heaven; and behold, they saw a Man descending out of heaven; and he was clothed in a white robe; and he came down and stood in the midst of them; and the eyes of the whole multitude were turned upon him, and they durst not open their mouths, even one to another, and wist not what it meant, for they thought it was an angel that had appeared unto them.
>
> 9 And it came to pass that he stretched forth his hand and spake unto the people, saying:
>
> 10 Behold, I am Jesus Christ, whom the prophets testified shall come into the world.

In 34 A.D. Jesus Christ appeared to the Nephites. By 36 A.D. the entire population had been converted to the Lord and a period of great peace and prosperity ensued. There were no classes and no poor and all things were had in common.

> 4 Nephi 1:7 And the Lord did prosper them exceedingly in the land; yea, insomuch that they did build cities again where there had been cities burned.
>
> 4 Nephi 1:10 And now, behold, it came to pass that the people of Nephi did wax strong, and did multiply exceedingly fast, and became an exceedingly fair and delightsome people.

Unfortunately, in 200 A.D. pride re-emerged, goods were no longer held in common and classes and disunity returned. The golden period was over at precisely the time indicated by the archaeological evidence. In 231 A.D. the people divided again into Nephites and Lamanites.

ROADS

The Hopewell were accomplished road builders. The most well-known of their roads is the Great Hopewell Road. "If Lepper's hypothesis is correct, a walled 'Road' some 65 yards wide and 56 miles long once ran directly from the octagonal 'observatory' at Newark to its astronomical counterpart at Chillicothe" (Charles River Editors, 2015, Chapter 5). As impressive as this road was the Hopewell road network is much more extensive than one road.

Mertz (2004) noted that a network of highways criss-crossing the Midwest was not built by the historic Indians who took advantage of them during the Indian wars but, rather, date back into prehistory. He wrote that "centuries ago a tangled web of highways threaded into this vast midwestern wilderness" (pp. 59-61). The most famous of these highways was known as the Mohawk Trail through central New York State. Other well-known trails running east west include the Kittanning Path, Nemacolin's Path and the Virginia Warrior's Path. Other trails of lesser significance "either connected or extended main trails" (p. 59-61). These trails included-the Lake Shore Trail, the Mahoning Trail, the Great Trail, The Scioto-Beaver-Monongahela Trail, the Venango Trail and the Cuyhoga-Muskingum Trail (Mertz, 2004, p. 59-61).

Who built these highways can be shown by the Scioto Trail which Romain (2000) noted "passed either directly through, or, quite literally, to within a few dozen feet of Portsmouth, Seal, Liberty, High Bank and Mound City enclosures" (p. 26). Subsidiary trails connected the Scioto Trail to the Cedar Bank and Circleville Earthworks (p. 26). As these trails connected major Hopewell centres it can be deduced that the Hopewell were responsible for their construction.

The Book of Mormon records that the Nephites constructed an extensive network of highways. "there were many highways cast up, and many roads made, which led from city to city, and from land to land, and from place to place" (3 Nephi 6:8).

SACRED ENCLOSURES

The most remarkable and intriguing of all Hopewell constructions were the earthworks known as sacred enclosures. The best known of these, near Newark, Ohio, is the largest and most precise geometric earthwork in the world, covering an area of four and a half square miles. According to Romain (2000) these earthworks prove that the Hopewell were "accomplished in astronomy, geometry, measuring, and counting (p. ix)." Astronomical observations over generations, advanced mathematical expertise and labour on a massive scale were required. But why did the Hopewell expend so much energy in their construction and what do they mean?

There is much speculation about the sacred enclosures but there are a few basic facts that are not in question. The enclosures were not places where populations lived, they were centres where people from the surrounding districts and perhaps from further afield, would gather from time to time to conduct special ceremonies or celebrations that united the community (Dancey & Pacheco, 1997, pp. 5-6). Their design included astronomical alignments with predominantly lunar but also solar cycles. Their construction included large squares, circles, octagons and avenues or walkways.

Back to the unanswered questions of why were the sacred enclosures built and what do they mean. Without first-hand experience, it is difficult to fully understand why ancient cultures did what they did, however, the Book of Mormon and modern sources can provide some useful perspectives.

LUNAR AND SOLAR ALIGNMENTS

As a starting point, we should remember that both the Lehites and Mulakites were Hebrews. Their calendar was a lunar calendar and keeping track of lunar movements was essential to observe the Law of Moses. The Law of Moses prescribed when the year started and when the major religious ceremonies were to be held.

> Yea, and the people did observe to keep the commandments of the Lord; and they were strict in observing the ordinances of God, according to the law of Moses; for they were taught to keep the law of Moses until it should be fulfilled (Alma 30:3).

The problem with lunar calendars is that it does not align with the solar calendar. There are approximately 12.4 lunar months in every solar year. To compensate for this difference the Jewish calendar usually has 12 months but occasionally adds an additional or 13th month.

To overcome this discrepancy early Hebrews used rather hit and miss methods. In Israel, the Sanhedrin would make observations of weather and crop and livestock conditions "and if these were not sufficiently advanced to be considered 'spring', then the Sanhedrin inserted an additional month into the calendar to make sure Pesach (Passover) would occur in the spring" (Rich, n.d.).

It was not until the fourth century in the old world that the Jews established "a fixed calendar based on mathematical and astronomical calculations" (Rich, n.d.). Perhaps they were preceded in their work by the Hopewell who, about two centuries earlier, calculated an 18.61 year lunar cycle to correlate with the solar cycle.

Boudinot (1816) observed that Indians in his day

> ... reckon their year after the manner of the Hebrews. They divide the year into spring, summer, autumn, or the falling of the leaf, and winter. Korak is their word for winter with the Cherokee Indians, as it is with the Hebrew.
>
> They begin their ecclesiastical year at the first appearance of the first new moon of the vernal equinox, according to the Ecclesiastical year of Moses (p. 164).

Aside from the regular Mosaic festivals, there were other reasons for the Nephites to keep an accurate calendar. For nearly six centuries the Nephites looked forward to the coming of Christ.

> 2 Nephi 25:19 For according to the words of the prophets, the Messiah cometh in six hundred years from the time that my father left Jerusalem; and according to the words of the prophets, and also the word of the angel of God, his name shall be Jesus Christ, the Son of God.

As the time approached in about 6 B.C. the prophet Samuel prophesied to the Nephites that

> Helaman 14:2 ... five years more cometh, and behold, then cometh the Son of God to redeem all those who shall believe on his name.
>
> 3 And behold, this will I give unto you for a sign at the time of his coming; for behold, there shall be great lights in heaven, insomuch that in the night before he cometh there shall be no darkness, insomuch that it shall appear unto man as if it was day.

As the day for the sign to be given drew close the unbelievers set a date that "all those who believed in those traditions should be put to death except the sign should come to pass, which had been given by Samuel the prophet" (3 Nephi 1:9). Just as the believers were "about to be destroyed" (3 Nephi 1:11) the sign was given.

The Nephites were diligent in recording their prophecies and their history. The text states that the sign of the Saviour's death was given "in the thirty and fourth year, in the first month, on the fourth day of the month" (3 Nephi 8:5). The author had confidence in his calendar and assured readers that the date was "according to our record, and we know our record to be true" (3 Nephi 8:1).

SYMBOLISM

Symbolism can be used to both hide or convey meaning. Symbols can be iconic or arbitrary. When a symbol is iconic a clue to its' meaning can be found in the symbol. When a symbol is arbitrary, only the users of that symbol understand its meaning. To help unravel the symbolism of the sacred enclosures the architecture of the Church of Jesus Christ of Latter-day Saints and other Christian churches has been drawn upon.

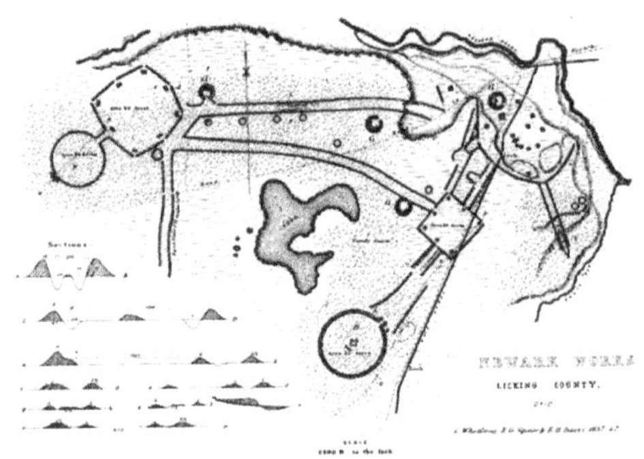

Source of Illustration:
Ancient Monuments of the Mississippi Valley,
Squier & Davis (1988)

Square

The square is a recurring shape in Hopewell enclosures. In church architecture it represents the earth. The shape relates to the four cardinal directions, north, south, east and west. The prophet

Isaiah wrote of gathering Israel from "the four corners of the earth" (Isaiah 11:12 and 2 Nephi 21:12).

Circle

Another shape used in all Hopewell enclosures is the circle. In church architecture the circle represents heaven. It is a shape which has no ending. When man stands on the earth and looks skyward the sky is a great circle.

The square and circle are used to represent earth and heaven in the architecture of Christian churches and temples of the Church of Jesus Christ of Latter-day Saints. In temples the circle drawn within a square represents the coming together of heaven and earth.

Octagon

Many Hopewell enclosures include an octagon. According to Reed (2011) the octagon, or squared circle, is a transitional shape between the square and the circle. The octagon is the connection between earth, represented by the square, and heaven, represented by the circle. It also symbolises Jesus Christ, who is the mediator between earth and heaven and connects them through his atoning sacrifice.

The shapes that were used by the Hopewell are used in temples today, the square representing the earth, the circle representing heaven, and the octagon representing the mediator, Jesus Christ. It is also notable that large circles in sacred enclosures had only one entrance, which may represent that there is only one way to heaven and that way was represented by the octagon, the mediator between earth and heaven, Jesus Christ.

AVENUES

According to Wagenaar (2005), processions may have been an important part of Jewish culture at the time that the Lehites left Jerusalem.

> The ancient Israelite New Year festival may, like the Babylonian New Year festivals, have been the occasion of a solemn procession. ... such procession may have been a regular part of the ancient Israelite cult in the late pre-exilic era (p. 110).

By the time of Christ the procession was an important preliminary of the Passover celebrations. Four days before the Passover, the priests would file out of the temple and weave their way through the streets of the city to the Damascus Gate. They then joined the swelling crowds waiting with palm fronds in hand. The High Priest selected the lamb to be sacrificed and entered the city through the Damascus gate. As they proceeded to the sacrifice, the crowds shouted "Hosanna to the Highest! Blessed is he who comes in the name of the Lord! Blessed be the kingdom of our father David that is coming! Hosanna in the Highest!" (Mark 11:9-10) (Mock, 2007).

Many of the Hopewell ceremonial sites included avenues lined by low walls. For example, The Newark earthworks had a series of at least six interconnected graded walkways flanked by low earthen walls. One of these avenues was two and a half miles long. We may ask, were the graded avenues built by the Hopewell used as venues for processions as preliminaries to Jewish and Christian celebrations held within the sacred enclosures?

SETTLEMENT PATTERN

Early researchers concentrated their efforts on the more impressive earthworks that we have previously discussed, however, in more recent years they have begun studying settlement patterns. Kozarek (1997) has identified several types of habitation sites in the Illinois region. All sites were permanently occupied and ranged from "large villages concentrated in major river valleys; other villages and homesteads along secondary streams; and small, isolated habitation sites (perhaps single extended families) in the uplands" (Kozarek, 1997, p. 132).

These settlements were often associated with mounds and were spread throughout the Mississippi valley. Mound centres, in the lower Illinois Valley were spaced about 12 miles apart. It is believed that mound centres in Wisconsin have similar spacing. Birmingham and Eisenberg (2000) believe that this spacing represents "the length of territories along the river that the centres served" (p. 92).

Proximity to natural resources was an important factor in the location of settlements. Each of the numerous sites are located near the five major resources that formed the basis of the Hopewellian diet. These resources were nuts and acorns, white-tailed deer, migratory waterfowl, various species of fish, and various seed plants (Struever 1968, as cited in Kingsley, 1981, p. 133).

The Book of Mormon refers to different types of settlement. The most frequently mentioned were lands and cities. The lands were the larger rural areas surrounding the cities and the cities were where the settlements were more concentrated with streets and houses, markets, synagogues and sometimes a temple. Towns and villages are also mentioned. The differences in the settlement densities is shown in the following verse:

> Alma 2:25 And they are upon our brethren in that land; and they are fleeing before them with their flocks, and their wives, and their children, towards our city; and except we make haste they obtain possession of our city, and our fathers, and our wives, and our children be slain.

There were numerous cities built throughout the Nephite territories.

> Mosiah 27:6 ... the people began to be very numerous, and began to scatter abroad upon the face of the earth, yea, on the north and on the south, on the east and on the west, building large cities and villages in all quarters of the land.

We should however be careful not to impose our own personal definition of a "city" upon the Nephites. Even today, different nations have different requirements to be met before an area can be designated as a city. In the United Kingdom it is the monarch that decides when a town becomes a city whereas in the USA a number of criteria must be met. When Nephi, the first leader of the Nephites, recounted a vision, he described Nazareth as a city (1 Nephi 11:13). At that time Nazareth was a small rural town with a population estimated at less than 400 persons. The Book of Mormon also records that Alma led a group of about 450 souls that built the city of Helam. The term city in the Nephite culture seems to refer more to function than size.

It is interesting to note that while archaeologists do accept that Hopewell settlements were permanent, they also suggest there were seasonal camps for the purpose of resource gathering. The Book of Mormon describes a population that lived in houses but possessed tents and could be mobile at short notice. When a gathering of the Nephites was called in about 124 B.C. the entire population gathered and camped in tents. When groups of Nephites were called to flee, or Nephite armies were called on to travel and when families moved to the treeless north, they used tents, just as American Indians in historic times, lived in tents.

TEMPLE MOUNDS

Another type of earthwork that the Hopewell built were the temple mounds. These had a rectangular base, sloping sides, and a flat top and were, therefore, sometimes referred to as truncated pyramids. They had graded ramps of earth leading from the ground to their tops. "To the discoverers it appeared probably that the flat topped mounds had once been platforms for Temples long ago destroyed by the elements" (Silverberg, 1970, p. 12).

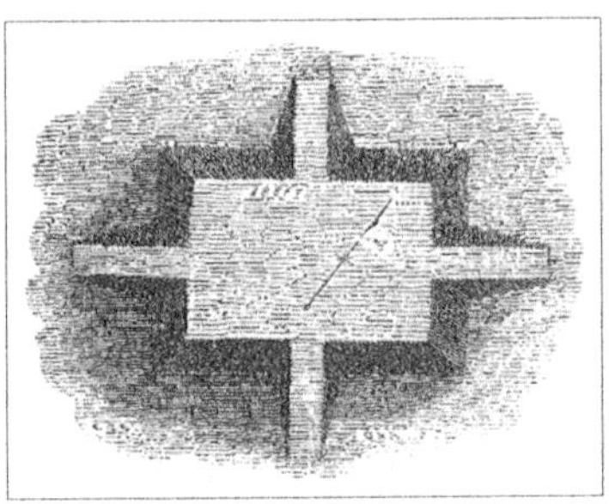

Source of Illustration:
Ancient Monuments of the Mississippi Valley,
Squier & Davis (1988)

None of the temple mounds contained burials. They were indeed the foundations for important buildings. What these buildings were can only be imagined but if any were for Hopewell temples it is interesting to compare them with Old Testament scripture.

> Exodus 20:26 Neither shalt thou go up by steps unto mine altar, that thy nakedness be not discovered thereon.
> Exodus 20:25 And if thou wilt make me an altar of stone, thou shalt not build it of hewn stone: for if thou lift up thy tool upon it, thou hast polluted it.

The Hopewell temple mounds had no steps and supported no carved altars. This is in stark contrast to the highly stylized stone temples and altars found in other parts of the world.

The Book of Mormon states that the Nephites built temples. Nephi built the first one "after the manner of the temple of Solomon save it were not built of so many precious things" (2 Nephi 5:16). When Christ appeared to the Nephites after his resurrection it was near the temple "which was in the land Bountiful" (3 Nephi 11:1).

TRADE

Archaeologists and others wishing to describe the extent of the Hopewell territories sometimes refer to their "interaction sphere" which stretched from the Great Lakes region in the north all the way to the Gulf of Mexico in the south. This depiction can be a little misleading. While for a period of time the Hopewell interacted with their southern neighbours, it must be made clear that the territory of the peoples we refer to as Hopewell did not extend any further south than Cairo, Illinois.

In the south a different people with a different culture, named Marksville, lived. "Virtually all cultural activity on a Hopewellian horizon from the latitude of Cairo, Illinois, south to the Gulf of Mexico is identified after this small town ... in Avoyelles Parish, Louisiana" (Toth, 1979, p. 188). As will be explained later, relations between the Hopewell and Marksville changed over time but, the Marksville never became Hopewell.

So, what was the sphere of influence? According to Smith (1979), shortly before the birth of Christ and continuing through the "first few centuries A.D." several cultures throughout most of the eastern woodland were trading or exchanging raw materials, manufactured goods and ideas (p. 181). "This network of contact and communication is labelled Hopewell or the Hopewellian interaction sphere" (Smith, 1979, p. 181).

This time period is significant from both an archaeological and Book of Mormon perspective. From a little before 1 A.D. to about 200 A.D. there was exchange and interaction between the Hopewell and their southern neighbours who have been termed Marksville. This interaction included the southern movement of Hopewellian style pottery from Illinois along with some cultural practices such as mortuary patterns. The Marksville did not become Hopewell but "adopted selectively and reinterpreted" the materials and practices from the north (Toth, 1979, p. 194). The contact and movement continued until 200 A.D. after which "there is little indication of sustained relations with the Illinois Valley" (Toth, 1979, p. 194).

What changed to allow the interaction during this period of time? For a little over 200 years friendly relations occurred between Hopewell and Marksville cultures but they maintained some cultural differences. After about 200 A.D. the interaction ceased as abruptly as it had started.

The answer to this question may be found in the Book of Mormon. Following successful missionary work by the Nephites amongst the Lamanites, "the Lamanites had become, the more part of them, a righteous people" (Helaman 6:1). Their hatred of their northern neighbours had been cast away.

> Helaman 6:8 And it came to pass that the Lamanites did also go whithersoever they would, whether it were among the Lamanites or among the Nephites; and thus they did have free intercourse one with another, to buy and to sell, and to get gain, according to their desire. (29 B.C.)

They traded freely for the first time. This period lasted until, as noted previously, in 231 A.D. the people divided again into Nephites and Lamanites.

> 4 Nephi 1:35 And now it came to pass in this year, yea, in the two hundred and thirty and first year, there was a great division among the people.

The period of time in which free trade between Lamanites and Nephites occurred, as recorded in the Book of Mormon, aligns precisely with the time period during which the Hopewell and Marksville Cultures traded, according to archaeological evidence.

WARS

The first chapter of this book touched upon the 19th century debate about the mounds and who created them. I wrote:

Many eyewitnesses could not believe that American Indians were capable of building such works. The Indian population was too small, their technology inadequate and their culture and nature did not seem suited to the organisation and labour required to perform such a mammoth undertaking. This belief was strengthened by the fact that when the Indians themselves were questioned they invariably denied any involvement in their construction or use and either knew nothing of their origins or referred to a previous race.

Most archaeologists today are scornful of any discussion of a vanquished race and label all such belief as 'the myth of the mound builders'. The campaign to sway opinion was initially led by officers of the Smithsonian Institute. As Silverberg (1970) wrote "the archaeologists who spoke for the Smithsonian were indeed able to impose their ideas on others-and they were not always correct" (p. 143).

According to Birmingham and Eisenberg (2000) the myth of the Lost Race was finally laid to rest by Cyrus Thomas's work *Report on the Mound Explorations of the Bureau of Ethnology* (pp. 31 & 33). His work contained uncollaborated accounts and serious inaccuracies but "made the important observation, supported by compelling evidence in the form of differing customs, constructions, and artifacts, that the various earthworks could not have been constructed by a single 'race' or group of Native American 'tribes and peoples'" (p. 33). This was the central argument that laid the myth to rest.

Thomas's central argument appears to be a syllogism. Different races or groups did construct earthworks, Americans construct earthworks today, but this does not preclude the possibility that one race or group that constructed earthworks in the past, namely, the Hopewell, were lost. The following information, drawn from archaeological books and papers, will outline the fate of the Hopewell.

SYSTEM OF DEFENCES

Previous sections have touched upon the fortified towns and places of refuge built by the Hopewell. Randall (1908) described lines of defence built by the Hopewell (p. 60) and Squier and Davis (1988) wrote of "system of defences extending from the sources of the Alleghany and Susquehanna in New York, diagonally across the country, through central and northern Ohio, to the Wabash" (p. 44). These structures were of sufficient number and size to prompt Squier and Davis (1988) to write,

"...it is clear that the contest was a protracted one, and that the race of the mounds were for a long period constantly exposed to attack" (p. 44).

To help understand the contest that the Hopewell were involved in, it may be helpful to look at the time periods in which their fortifications were constructed.

- Fort Miami in Hamilton County. A radio carbon date obtained from a sample of charred wood from the original sub-embankment surface "yielded a date of AD 270 plus or minus 130" (Prufer, 1997, p. 316).

- Shawnee Lookout Park, Hamilton County, Ohio. "A radio carbon dated sample of burned earth suggests that construction of the fort began around AD 270" (Woodward & McDonald, 1986, p. 99).

- Pollock earthworks in Greene County. "The period of fence and stockade building took place at one time, quickly, sometime between AD 130 and AD 228" (Woodward & McDonald, 2001, p. 153).

- "At present, archaeological evidence indicates that Fort Ancient was constructed in stages between about 100 B.C. and AD 300" (Woodward & McDonald, 2001, p. 243).

- Old Stone Fort. "Available radiocarbon dates suggest that the enclosure was begun early in the first century A.D. ... but that most of the wall construction took place in the period of A.D. 200 to A.D. 400" (Butler, 1979, p. 150).

Accurately dating sites is difficult and we do not have a lot of dates to work from. There is significant evidence that fortifications were also built at other times. Prufer (1997) noted that "all southern Ohio hilltop enclosures that had been tested and reported on are characterised by at least two chronologically distinct construction phases" (p. 319). The relevant point to be drawn from the data we have is that some time after the Hopewell passed that pivotal year of 200 A.D., which marked the end of their golden period and the end of trade with their southern neighbours, the Hopewell began furiously building and strengthening their fortifications. The construction did not commence straight away, perhaps not until 270 A.D., but at some point in time their situation deteriorated to the point where they felt such preparations were necessary.

AGGREGATION

As time passed, the Hopewell lost the freedom to spread over their prime lands as they had previously done. As the threat increased it became necessary for communities to gather together for security. Abrams (2009) recognised that settlement data and chronology are very limited but was prepared to state that "a physical restructuring of Ohio communities on a local level following c. A.D. 300-400 appears to have occurred" (p. 186).

Abrams (2009) drew on data collected from sites such as the Water Plant site and Swinehart Village site to proffer a view that smaller Hopewellian sites were abandoned and populations moved into larger, defensive, often fortified, communities. These communities were not in the old Hopewellian heartlands but were on the margins of the lands used by earlier Hopewell. In some areas "a large-scale abandonment of the valley is evidenced" (Abrams, 2009, pp. 186-187).

Charles River Editors (2015) used this aggregation of the Hopewell population from smaller communities to "larger communities fortified with walls and ditches" as evidence to support the belief of some scholars that an "inter-regional war was the cause of their demise" (Charles River Editors, 2015, Chapter 6).

CESSATION OF NON-ESSENTIAL ACTIVITIES

As with any war effort, non-essential activities were curtailed and all energies were focused on addressing the more pressing need. The construction of large burial mounds and ceremonial sites ceased. Archaeological records align with the Book of Mormon. For example,

- "Beginning ca. A.D. 250 the construction of burial mounds and other earthworks ceased as a cultural practice throughout the Ohio Valley. The Hocking Valley is no exception" (Abrams & Freter, 2005, p. 189).
- "All that is known for sure is that the monumental mound and earthwork construction associated with this flamboyant period of north American pre-history ceased sometime between AD 200 and AD 400" (Birmingham & Eisenberg, 2000, p. 98).
- "Earthwork construction appears to have terminated between AD300-400" (Dancey & Pacheco, 1997, p. 17).

ABANDONMENT

The next action of the Hopewell was, in around 350 A.D., to abandon their lands in the Mississippi valley.

- At the end of this phase, sometime around *cal* A.D. 350, the gradual process of Hopewellian collapse in the American Bottom is suspended by the apparent abandonment of the area (Fortier, 2001, p. 191).
- Although speculative, these new intercommunity tensions emerged after A.D. 350, just when the local and regional Middle Woodland ceremonial and mortuary centres were apparently abandoned (Schweikart, 2008, p. 124).
- There is an apparent abandonment of Mississippi River floodplain locales during the period *cal* A.D. 400-650 that is inexplicable without invoking ruminations of an environmental nature, such as, Was the Garden of Eden under water? (Fortier, 2001, p. 178)

RETREAT

When Squier and Davis completed their inventory of earthworks, they wrote that all those constructed in northern Ohio were built for military purposes. Although they did not have direct evidence, they felt it safe to imagine that the earthen walls they found there were once topped with wooden palisades, and posts and gates protected the passages left in the embankments and ditches. With a complete absence of ceremonial sites and burial mounds it appeared to them that the population in this area was so focused on war that they must have been a different people.

The careful positioning of the northern works was also discussed. They were all adjoining water and were on the highest lands in the immediate vicinity. Those along the shoreline of Lake Erie in Ohio could only have been intended for defence. Together, the works formed a defensive line "from Conneaut to Toledo, at a distance from three to five miles from the lake, and all stand upon or near the principle rivers" (Squier & Davis, 1988, p. 41). Larger works were constructed in less favourable positions, such as lower lands, but compensated with the strength of their construction.

The cordon of defences was not limited to northern Ohio but stretched eastward into New York and Pennsylvania. Squier and Davis (1988) reported similar works were found further to the north and to the east, "extending to the Genesee river and its tributaries in New York, and even to the headwaters of the Susquehanna river in Pennsylvania" (p. 42).

The defences also extended westward. In 1817, army engineer Stephen H. Long, provided his assessment of earthworks in Wisconsin. He described numerous and extensive ancient earthworks above the mouth of the Ouisconsin which "consist of ridges or parapets of earth and mounds variously disposed as to conform to the nature of the ground they are intended to fortify" (Birmingham & Eisenberg, 2000, p. 4).

Squier and Davis (1988) also noted that the fortifications in the north were of different design than those in the south. Instead of single earthen walls, northern fortifications had multiple walls with an external ditch. The exterior walls sometimes had passages or 'sally ports' but there were no gaps in the inner wall. Between the inner and outer walls were usually spaces that were large enough to protect large numbers of fighting men (p. 41).

Summing up their deliberations on the northern earthworks, Squier and Davis (1988) stated that the builders of the north "were much engaged in offensive, or defensive wars. At the south, on the other hand, agriculture and religion seemed to have chiefly occupied the attention of the ancient people" (p. 41). They concluded that, in their opinions, the works stretching across the north:

formed a well-occupied line, constructed either to protect the advance of a nation landing from the lake and moving southward for conquest; or, a line of resistance for a people inhabiting these shores, and pressed upon by their southern neighbours (Squier & Davis, 1988, p. 41).

Squier and Davis did not have the advantages of the years of archaeological research that we have today and, hence, their conjecture that the northern areas may have been inhabited by a different people. We now know that it was the Hopewell who built the northern fortifications. Squier and Davis were accurate in their views regarding it being a line of resistance however, the changes in behaviour the population displayed were not because they were of a different race but because they were in retreat, locked in a desperate life or death struggle with their southern neighbours.

END

In around 400 A.D. the Hopewell culture disappeared from the American landscape. In the words of Dancey and Pacheco (1997) "... the entire Hopewell system evolves and then ultimately disappears from the record (near the Late Hopewell/Early Late Woodland boundary ca. AD 400)" (p. 59).

Why did the Hopewell Culture disappear? According to Abrams (2009), it was not due to economic failure, overpopulation or environmental change. Abrams reasoned that economic failure is often associated with overpopulation which causes environmental degradation. This leads to increased death rates and the movement of populations out of the affected areas. However, the evidence of such changes has not been found. There are no signs of disease epidemics, increased death rates or environmental collapses (Abrams, 2009, pp. 186-187).

What then caused the Hopewellian disappearance? Abrams listed the views of a number of researchers as follows:

- Braun (1986-7, as cited in Abrams, 2009, pp. 186-187) logically inferred that if risks prompted regional integration, then success at overcoming those risks prompted the cessation of institutions of integration, and
- Dunnell and Greenlee (1999) argued that earthwork construction and long-distance exchange were eventually selected against by

Hopewellian societies when reclassified as dysfunctional, or wasteful (as cited in Abrams, 2009, pp. 186-187).

Abrams concluded that it is obvious that we need more data if we are going to better explain "the regional variability of the transformation of Hopewellian societies if we are going to then explain this important historical process" (Abrams, 2009, pp. 186-187).

More data is better, but the information we already have begs the question, why did Abrams not include warfare in his list of possible causes for the cessation of the Hopewell Culture? The Hopewell expended massive labours to build fortifications, they abandoned their smaller communities in favour of larger, better defended centres, they ceased the construction of burial mounds and ceremonial sites, they physically abandoned the Mississippi valley which had been their home for centuries, and they constructed a line of heavy fortifications in the north. Is the notion that they had overcome the risks that they faced and decided that their activities over all those centuries were dysfunctional or wasteful the most obvious explanation? Why was warfare not considered?

There are still some archaeologists who refuse to contemplate warfare as a possible cause of the Hopewell demise. I will discuss the reasons for this in a later chapter. Regardless of their reticence, the archaeological record describes a sequence of events that shares many similarities with the Book of Mormon account which culminated in the military defeat and destruction of the Nephite people.

The Book of Mormon provides the following time line of the end of the Nephite nation.

> 4 Nephi 1:35 And now it came to pass in this year, yea, in the two hundred and thirty and first year, there was a great division among the people. (231 A.D.)

Note that this timing agrees with the timing of the cessation of trade between the Hopewell and Marksville cultures.

> 2:1 And it came to pass in that same year there began to be a war again between the Nephites and the Lamanites. (326 A.D.)
>
> 2:3 And it came to pass that in the three hundred and twenty and seventh year the Lamanites did come upon us with exceedingly great power, insomuch that they did frighten my armies;

therefore they would not fight, and they began to retreat towards the north countries. (327 A.D.)

Note the northward retreat of the Nephite armies.

2:4 And it came to pass that we did come to the city of Angola, and we did take possession of the city, and make preparations to defend ourselves against the Lamanites. And it came to pass that we did fortify the city with our might; but notwithstanding all our fortifications the Lamanites did come upon us and did drive us out of the city.

2:5 And they did also drive us forth out of the land of David.

2:6 And we marched forth and came to the land of Joshua, which was in the borders west by the seashore.

2:7 And it came to pass that we did gather in our people as fast as it were possible, that we might get them together in one body.

2:16 And it came to pass that in the three hundred and forty and fifth year the Nephites did begin to flee before the Lamanites; and they were pursued until they came even to the land of Jashon, before it was possible to stop them in their retreat. (345 A.D.)

2:20 And it came to pass that in this year the people of Nephi again were hunted and driven. And it came to pass that we were driven forth until we had come northward to the land which was called Shem.

2:21 And it came to pass that we did fortify the city of Shem, and we did gather in our people as much as it were possible, that perhaps we might save them from destruction.

2:22 And it came to pass in the three hundred and forty and sixth year they began to come upon us again. (346 A.D.)

2:28 And the three hundred and forty and ninth year had passed away. And in the three hundred and fiftieth year we made a treaty with the Lamanites and the robbers of Gadianton, in which we did get the lands of our inheritance divided. (350 A.D.)

2:29 And the Lamanites did give unto us the land northward, yea, even to the narrow passage which led into the land southward. And we did give unto the Lamanites all the land southward.

4 Nephi 3:5 And it came to pass that I did cause my people that they should gather themselves together at the land Desolation, to a city which was in the borders, by the narrow pass which led into the land southward.

3:6 And there we did place our armies, that we might stop the armies of the Lamanites, that they might not get possession of any of our lands; therefore we did fortify against them with all our force.

Note that the Nephites fortified the lands around the narrow pass between the seas which led into the land northward.

3:7 And it came to pass that in the three hundred and sixty and first year the Lamanites did come down to the city of Desolation to battle against us; and it came to pass that in that year we did beat them, insomuch that they did return to their own lands again. (361 A.D.)

3:8 And in the three hundred and sixty and second year they did come down again to battle. And we did beat them again, and did slay a great number of them, and their dead were cast into the sea. (362 A.D.)

4 Nephi 4:1 And now it came to pass that in the three hundred and sixty and third year the Nephites did go up with their armies to battle against the Lamanites, out of the land Desolation. (363 A.D.)

4:2 And it came to pass that the armies of the Nephites were driven back again to the land of Desolation. And while they were yet weary, a fresh army of the Lamanites did come upon them; and they had a sore battle, insomuch that the Lamanites did take possession of the city Desolation, and did slay many of the Nephites, and did take many prisoners.

4:3 And the remainder did flee and join the inhabitants of the city Teancum. Now the city Teancum lay in the borders by the seashore; and it was also near the city Desolation.

4:4 And it was because the armies of the Nephites went up unto the Lamanites that they began to be smitten; for were it not for that, the Lamanites could have had no power over them.

Note that the Nephites had constructed a line of defences around the seas that would have been sufficiently strong to defend their northern lands from the Lamanites if they had not themselves ventured beyond their fortifications seeking revenge on their enemies.

4:7 And it came to pass in the three hundred and sixty and fourth year the Lamanites did come against the city Teancum, that they might take possession of the city Teancum also.

4:8 And it came to pass that they were repulsed and driven back by the Nephites. And when the Nephites saw that they had driven the Lamanites they did again boast of their own strength; and they went forth in their own might, and took possession again of the city Desolation. (366 A.D.)

4:13 And it came to pass that the Lamanites did take possession of the city Desolation, and this because their number did exceed the number of the Nephites.

4:14 And they did also march forward against the city Teancum, and did drive the inhabitants forth out of her, and did take many prisoners both women and children, and did offer them up as sacrifices unto their idol gods.

4:15 And it came to pass that in the three hundred and sixty and seventh year, the Nephites being angry because the Lamanites had sacrificed their women and their children, that they did go against the Lamanites with exceedingly great anger, insomuch that they did beat again the Lamanites, and drive them out of their lands. (367 A.D.)

4:16 And the Lamanites did not come again against the Nephites until the three hundred and seventy and fifth year. (375 A.D.)

4:17 And in this year they did come down against the Nephites with all their powers; and they were not numbered because of the greatness of their number.

4:18 And from this time forth did the Nephites gain no power over the Lamanites, but began to be swept off by them even as a dew before the sun.

4:19 And it came to pass that the Lamanites did come down against the city Desolation; and there was an exceedingly sore battle fought in the land Desolation, in the which they did beat the Nephites.

4:20 And they fled again from before them, and they came to the city Boaz; and there they did stand against the Lamanites with exceeding boldness, insomuch that the Lamanites did not beat them until they had come again the second time.

4:21 And when they had come the second time, the Nephites were driven and slaughtered with an exceedingly great slaughter; their women and their children were again sacrificed unto idols.

4:22 And it came to pass that the Nephites did again flee from before them, taking all the inhabitants with them, both in towns and villages.

4 Nephi 5:3 And it came to pass that the Lamanites did come against us as we had fled to the city of Jordan; but behold, they were driven back that they did not take the city at that time.

5:4 And it came to pass that they came against us again, and we did maintain the city. And there were also other cities which were maintained by the Nephites, which strongholds did cut them off that they could not get into the country which lay before us, to destroy the inhabitants of our land.

5:5 But it came to pass that whatsoever lands we had passed by, and the inhabitants thereof were not gathered in, were destroyed by the Lamanites, and their towns, and villages, and cities were burned with fire; and thus three hundred and seventy and nine years passed away. (379 A.D.)

5:6 And it came to pass that in the three hundred and eightieth year the Lamanites did come again against us to battle, and we did stand against them boldly; but it was all in vain, for so great were their numbers that they did tread the people of the Nephites under their feet.

5:7 And it came to pass that we did again take to flight, and those whose flight was swifter than the Lamanites' did escape, and those whose flight did not exceed the Lamanites' were swept down and destroyed.

4 Nephi 6:4 And it came to pass that we did march forth to the land of Cumorah, and we did pitch our tents around about the hill Cumorah; and it was in a land of many waters, rivers, and fountains; and here we had hope to gain advantage over the Lamanites.

6:5 And when three hundred and eighty and four years had passed away, we had gathered in all the remainder of our people unto the land of Cumorah. (384 A.D.)

6:7 And it came to pass that my people, with their wives and their children, did now behold the armies of the Lamanites marching towards them ...

6:8 And it came to pass that they came to battle against us, and every soul was filled with terror because of the greatness of their numbers.

6:9 And it came to pass that they did fall upon my people with the sword, and with the bow, and with the arrow, and with the ax, and with all manner of weapons of war.

6:10 And it came to pass that my men were hewn down ...

The final war began in 326 A.D. The Nephites retreated northwards. Eventually they were able to construct sufficient fortification to prevent any further advances by the Lamanites. When they left the protection of their defences they were defeated and their actions allowed their defences to be breached. Over the years that followed the Nephites were pursued and beaten until in 384 A.D. the final battle was fought which ended in the extinction of the Nephite nation. The description and timing of these events aligns with the archaeological record and the northern fortifications, described by Squier and Davis and others, stand as a silent testimony to what took place. Perhaps the conclusions of the early explorers and archaeologists, regarding warfare and defences, were not as ill-considered as later archaeologists would have us believe.

WEAPONS

An early researcher, Atwater, found possible evidence of a metal sword near two skeletons in a mound in Circleville. Beside the skeletons was a rusted metal plate which "seemed to resemble cast iron" along with "a handle made of elk's horn, at one end of which were traces of rust indicating the onetime presence of the iron blade of a sword or large knife" (Silverberg, 1970, p. 55).

The possibility of iron weapons has been supported by Marder (2005), who believes that ancient furnaces have been found. He stated that 130 iron or copper-smelting furnaces have been discovered in Ohio, centred in Ross County and "a few in Virginia, Georgia, and Kentucky, and as far west as Arizona and New Mexico" (Marder, 2005, p. 27-28).

The Book of Mormon describes the use of metal swords by the Nephites.

> Alma 2:12 ... they did arm themselves with swords, and with cimeters, and with bows, and with arrows, and with stones, and with slings, and with all manner of weapons of war, of every kind.

The Book also states that they used bows and arrows. In one passage however, it mentions casting stones and arrows. This may refer to the use of the atlatl, a weapon of the Hopewell that was similar in function to the woomera, used by Indigenous Australians, and enabled the user to throw their projectiles a much greater distance.

> Alma 49:4 But behold, how great was their disappointment; for behold, the Nephites had dug up a ridge of earth round about them, which was so high that the Lamanites could not cast their stones and their arrows at them that they might take effect ...

It is notable that Nephite armour was described as breastplates and headplates, not helmets. The Hopewell wore copper breastplates and copper headplates, the latter of which consisted of a strip of metal that followed the centre of the skull from the top of the forehead to the back of the head, unlike a helmet which we are all familiar with.

> Alma 49:24 ... they were shielded by their shields, and their breastplates, and their head-plates ...

WRITING

Writing is the single most controversial topic in relation to the Hopewell Culture. The current attitude of many archaeologists is demonstrated by Woodward and McDonald (1986). Their introduction to the Adena and Hopewell Cultures is as follows: "The mound builders did not leave written records in books of their origins, achievements and beliefs, as do people of true civilisations" (p. 5). Archaeologists accept that the Hopewell were capable of constructing massive earthworks that were geometrically perfect and tracked the movement of the moon through an 18.61 year cycle that could only be determined by generations of observations but they will not concede that they could write. How is this possible?

The Hopewell left more than enough samples of their writings. However, to date, much of the academic world, led by the Smithsonian Institute, has decried every such example as fraudulent even when discovered by staff of the Smithsonian Institute, or in other cases, the items have disappeared. Here are some examples.

- The Newark/Ohio Decalogue was discovered in a burial mound in November 1860 by David Wyrick of Newark Ohio. The mound was near the famous Newark earthworks. The stone is a finely grained black stone and was discovered fitted perfectly inside a smooth, hollowed out, sandstone box. The black stone bears a beautifully crafted carving showing a bearded man with fine features and a mild expression on the front wearing a turban and flowing robe. He is either holding a tablet or wearing a breastplate. The name Moses is inscribed over his head. A condensed version of the Ten Commandments is inscribed on all sides of the stone. All the writings are in a "peculiar form of post-exilic, square Hebrew letters" (Marder, 2005, p.48).

The Decalogue Stone

- Another stone inscribed with Hebrew letters was also found by Wyrick in June 1860. "This stone dated to circa 100 B.C. to 500 A.D" (Marder, 2005, p. 48).
- The Grave Creek tablet was discovered in a mound in the Grave Creek earthworks in June 1848 by Jesse and Abelard Tomlinson. It was an oval disc made of white sandstone and was one and a half inches in diameter and three quarters of an inch thick. On it "were inscribed three lines in an unknown alphabet" (Silverberg, 1970, p. 76).
- The Bat Creek stone was found by the Smithsonian Bureau of Ethnology in 1889 in a burial mound in Tennessee. The stone was engraved with letters which have been identified as possibly Paleo-Hebrew of the first or second century A.D. (Bat Creek Inscription, (n.d.)
- "In the interval between 1837 and 1875, scores of other inscribed stones turned up in areas from the Eastern seaboard to as far west as the Mississippi River" (Mertz, 2004, p. 4).
- The Standing Stone was previously located at Huntingdon, on the Juniata River, due east of Altoona. The spot was previously named Standing Stone after an ancient stone obelisk which was 14 feet high and broad at the base but tapering to six or seven inches at the apex. The obelisk was covered on all four sides with "undecipherable letters which early explorers believed to be Egyptian hieroglyphics" (Mertz, 2004, pp. 62-63). It was conjectured that the marker may have been a directional guide for ancient travellers. It was a well-known landmark and is frequently mentioned in Pennsylvanian archives. During the Revolutionary War, Colonial soldiers would gather at its base. "The original stone mysteriously disappeared sometime between 1754 and 1755" (Mertz, 2004, pp. 62-63).
- "Another towering ancient stone once occupied the site of present-day Lancaster, Ohio-in an area choked with ancient mounds and fortifications" (Mertz, 2004, pp. 62-63).

Writings have been found buried in ancient mounds and on towering markers. How many more were found by people unwilling to face the persecution, ridicule and slander heaped upon discoverers by the academic establishment will never be known. There is enough evidence to argue that the Hopewell wrote in both hieroglyphics and in ancient Hebrew.

Such writing is still in use today among some native American tribes. Marder (2005) wrote that the Algonquian speaking tribes in the northeast wrote vertically on tree bark using Egyptian style hieroglyphic script symbols. He also stated that some tribes from the southwestern plains write from right to left using a pictorial style of writing (Marder, 2005, p. 57).

The Book of Mormon states that the Nephites wrote in hieroglyphics, also known as reformed Egyptian and in Hebrew.

> Mormon 9:32 And now, behold, we have written this record according to our knowledge, in the characters which are called among us the reformed Egyptian, being handed down and altered by us, according to our manner of speech.
>
> Mormon 9:33 And if our plates had been sufficiently large we should have written in Hebrew; but the Hebrew hath been altered by us also; and if we could have written in Hebrew, behold, ye would have had no imperfection in our record.

While it may be a circular argument, if the Book of Mormon is a true record then the mound builders do meet the criteria for 'true civilisations' suggested by Woodward and McDonald (1986) in that they did leave written records in books of their origins, achievements and beliefs.

5

Historic Period (1492 A.D.–today)

FOLLOWING THE DESTRUCTION of the Nephite nation the victorious Lamanites, who included Nephite defectors, were left to possess the former Nephite lands. However, things did not go smoothly for the Lamanites. With the defeat of the common enemy the Lamanites turned on each other. Seventeen years after the battle at Cumorah, Moroni, the last Nephite survivor, wrote "the Lamanites are at war one with another; and the whole face of this land is one continual round of murder and bloodshed; and no one knoweth the end of the war" (Mormon 8:8).

We have no record of what happened following Moroni's death. When Europeans arrived, they found a land inhabited by many different tribes speaking many different languages. The following section focuses on the tribes of the Algonquian and Iroquois language groups that inhabited the Atlantic Coast, Great Lakes region, eastern Canada and some midwestern states when European settlers arrived.

HEBREW CONNECTIONS

In the early days of the mound builder debate, people such as Boudinot (1816), drew attention to similarities between some native American tribes and Hebrew peoples in relation to language, calendar, and religious beliefs. Some modern archaeologists continue in this belief. Charles River Editors (2015) state that modern Ojibway Native Americans (Algonquian language group) "trace their culture to the Lost Tribes" of Israel (Chapter 1). They cite Ojibway scholar William W. Warren who wrote that the Ojibway are "descendants of the lost tribes of Israel, or they had, in some former era, a close contact and intercourse with the Hebrews, imbibing from them their beliefs and customs and the traditions of their patriarchs" (Charles River Editors, 2015, Chapter 1).

Unlike many native peoples, these native Americans are monotheistic, believing in a "Great Spirit". It is significant to note that the Book of Mormon records the Lamanite belief in a Great Spirit.

> Now this was the tradition of Lamoni, which he had received from his father, that there was a Great Spirit. Notwithstanding they believed in a Great Spirit, they supposed that whatsoever they did was right; nevertheless, Lamoni began to fear exceedingly, with fear lest he had done wrong in slaying his servants (Alma 18:5);

THE PATH OF SOULS

One of the most common beliefs among native American tribes in historic times relates to the journey of the soul after death. According to this belief, after death the soul leaves the body and travels along a path towards the Land of the Dead. The path is known by various names such as *the Path of Souls, Pathway of Departed Spirits, Spirit's Path, Spirit Way, Spirit's Road* or *Ghost's Road* (Miller & Lipsett, 1998). Along this path the departed soul faces "challenges, tests, or obstacles" (Lankford, 2007).

The Path is often associated with the Milky Way in the night sky. When the soul reaches the Great Rift, at the star Deneb, it encounters a test. A creature which may take the form of an old woman, or a dog, or some other adversary, blocks the way. If the soul "successfully meets the challenge or is deemed worthy" it is allowed to cross the rift to the path that leads to the Land of the Dead where it is reunited with its worthy ancestors. If it fails the test it is sent along the other path to "oblivion, or other bad outcomes" (Romain, 2015, p. 28)

While it may be coincidental, there are parallels between this native American belief and two verses from the Book of Mormon.

> Helaman 3:29 Yea, we see that whosoever will may lay hold upon the word of God, which is quick and powerful, which shall divide asunder all the cunning and the snares and the wiles of the devil, and lead the man of Christ in a strait and narrow course across that everlasting gulf of misery which is prepared to engulf the wicked-
>
> 30 And land their souls, yea, their immortal souls, at the right hand of God in the kingdom of heaven, to sit down with

Abraham, and Isaac, and with Jacob, and with all our holy fathers, to go no more out.

MISSIONS TO THE LAMANITES

Early in the nineteenth century, following the publication of the Book of Mormon, Joseph Smith was commanded to send missionaries to the descendants of the Lamanites.

> And now, behold, I say unto you that you shall go unto the Lamanites and preach my gospel unto them; and inasmuch as they receive thy teachings thou shalt cause my church to be established among them; and thou shalt have revelations, but write them not by way of commandment (D&C 28:8).
>
> And now concerning my servant Parley P. Pratt, behold, I say unto him that as I live I will that he shall declare my gospel and learn of me, and be meek and lowly of heart.
>
> And that which I have appointed unto him is that he shall go with my servants, Oliver Cowdery and Peter Whitmer, Jun., into the wilderness among the Lamanites (D&C 32:1-2).

In response to this direction from the Lord, missionaries went to the Cattaraugus Indians (Iroquois) near Buffalo, New York; Wyandots (Iroquois), Ohio; and Delawares (Algonquian), West of Missouri. At another time, Joseph Smith himself taught members of the Sac and Fox (Algonquian) tribes. Elder B. H. Roberts gave an account of the visit.

Joseph Smith met the Indians at a grove of trees and spoke to them, at some length, "upon what the Lord had revealed to him concerning their forefathers" (A Comprehensive History of the Church of Jesus Christ of Latter-day Saints, 2:88-89). The link between the Sac and Fox tribe and the Book of Mormon was unambiguous. He went on to explain the promises in the Book of Mormon that related specifically to them.

Elder Roberts mused on the reaction of the Indians to Joseph's words. "How their hearts must have glowed as they listened to the prophet relate the story of their forefathers—their rise and fall; and the promises held out to them of redemption from their fallen state!" (A Comprehensive History of the Church of Jesus Christ of Latter-day Saints, 2:88-89).

Elder Roberts went on to relate the response made by the Indian spokesman, Keokuk. Keokuk told Joseph that he had in his wigwam the

Book of Mormon that Joseph had given him some months earlier. He stated his belief that Joseph was a great and good man and declared that, though rough looking "I am a son of the Great Spirit. I have heard your advice. We intend to quit fighting, and follow the good talk you have given us" (A Comprehensive History of the Church of Jesus Christ of Latter-day Saints, 2:88–89).

From the preceding account of the meeting between Joseph Smith and the Sac and Fox tribe it is clear that Joseph Smith believed that they were descendants of the Lamanites of the Book of Mormon.

DNA EVIDENCE

In the preface of this book I wrote that Church opponents had used DNA evidence to argue against the Book of Mormon based on the Asian origins of populations in Central and South America. Unfortunately, this research did not include the North American Indians that Joseph Smith identified as descendants of the Lamanites.

Marder (2005) explained that in 1997 another mtDNA group, called X, was discovered in both living Indians and prehistoric remains. This was a significant finding on at least two fronts.

- The new X mtDNA is not of Asian origins but is linked to Europe and the Middle East. Haplogroup X is found in Israel, Finland, Bulgaria, the Basque region of Spain, and in extreme south Siberia.
- Who the X mtDNA was found amongst was also significant. It is found in "small numbers among the northern Indian tribes of the Na-Dene speaking Navajo, Yakama (Yakima), Nuu-Chah-Nulth, the Sioux" (Marder, 2005, p. 16). The important finding is that it is found "in larger numbers among the Ojibway (Ojibwe, Chippewa), Iroquois and Oneida tribes, and in ancient remains in Illinois near Ohio and near the Great Lakes" (Marder, 2005, p. 16). Marder stated that geneticists are certain that this DNA has not come into the population through modern intermarriages but is very ancient. Please note that Ojibway is part of the Algonquian language group and Oneida is part of the Iroquois language group.

Unlike Central and South America populations which carry Haplogroups A, B, C, and D of Asian origin the North American tribes carry DNA of Haplogroup X which is believed to originate in the Middle East and is most common in the Galilee region of Israel. The X mtDNA is particularly common in Algonquian and Iroquois tribes.

Most anthropological geneticists currently reject the notion that the presence of X mtDNA in North American Indian populations supports pre-Columbian trans-Atlantic migration. The North American mtDNA is a unique strain called X2a which, though it descended from X, has not been found anywhere else in the world. The geneticists prefer to believe that the X2a variant developed in a population that was isolated in Siberia or Beringia for an extended period of time even though they have no solid evidence to support this conclusion. Placing the development of the X2a mtDNA in Siberia or Beringia fits in with their pre-existing notions of migration through those regions to America. However, Raff and Bolnick (2015) concede that:

> Should credible evidence of direct gene flow from an ancient Solutrean (or Middle Eastern) population be found within ancient Native American genomes, it would require the field to reassess the "Beringian only" model of prehistoric Native American migration.

While it is not possible at this time to conclude that DNA evidence proves that the Book of Mormon is true, it can be categorically stated that, contrary to the claims made by critics, DNA evidence does not disprove the Book but leaves open the possibility of trans-Atlantic migration from the Middle-east to North America.

6

The Influencers

OVER THE PAST 150 YEARS archaeologists, anthropologists, and ethnologists have continued to study the mounds of the Mississippi Valley. Yet despite these efforts Birmingham and Eisenberg are prepared to state that "the mounds continue to be a mystery because perceptions and interpretations about them continually change—molded and limited by the social and scientific climates of the times" (Birmingham & Eisenberg, 2000, p. 12).

Birmingham and Eisenberg (2000) should be applauded for their candid honesty. Despite confident statements made by some archaeologists the mounds continue to be a mystery. Thanks are also warranted for acknowledging the limitations imposed by social and scientific climates. These climates will now be discussed.

In the early nineteenth century many notable people, including the President of the United States, concluded that the mounds were built by a race that had long since vanished from the landscape. Their conclusions seemed rational and logical and had the support of the native American tribes that then inhabited the land. However, a small group of men, led by John Wesley Powell, began a concerted effort to change public opinion.

John Wesley Powell (1881)
Sourced from Wikipedia

What was the motivation of John Wesley Powell? His adult exploits are well documented but there is little written about his childhood. We know that his father was an itinerant preacher for the Methodist church and his mother was a missionary. His father immigrated from England to America in 1830 into the midst of an unusual set of circumstances.

Central and western New York state in the early nineteenth century were known as the burned-over district because of the religious revivals and excitement that had permeated that area. Adding to the uproar was the news that a young farm boy, by the name of Joseph Smith Jnr., had found gold plates containing a history of the ancient inhabitants. In 1830 the Church of Jesus Christ of Latter-day Saints was established in Fayette, New York, the same year that Powell's father arrived. John Wesley Powell was born in 1834 in Mount Morris, New York, about 65 miles from Fayette and 50 miles from the old Smith home at Palmyra. It is unclear how growing up at that time, in that area, in a very Methodist household impacted Powell.

The social climate permeating Powell's world was also influenced by the "Monroe Doctrine" and "Manifest Destiny". The Monroe Doctrine "asserted that the New World and the Old World were to remain distinctly separate spheres of influence, for they were composed of entirely separate and independent nations" (Monroe Doctrine, n.d.). Manifest Destiny "was a widely held belief in the United States that its settlers were destined to expand across North America" (Manifest Destiny, n.d.).

How these influences coalesced was explained by Scherz (2001). Scherz wrote that all artefacts that point to a connection with the Old World before Columbus suffer the same treatment from academic authorities. In the 1800s, North American archaeologists and authorities appear to have agreed on a policy of disallowing any links between Europe and America before Columbus. He likened this policy to the Monroe Doctrine which reserves America for Americans and warns against any attempted interference from Europe.

Scherz traced the emergence of the policy to John Wesley Powell. Powell was an officer in the civil war and an explorer however his scholarship was questionable. He had studied at University for seven years but was unable to obtain a degree. Despite this he was appointed to the position of director of the Bureau of Ethnology at the Smithsonian Institution and was elected a member of the American Antiquarian Society.

Powell "was extremely effective in putting an end to the mound builder controversy that was then creating such excitement in the land" (Scherz, 2001, pp. 1-5). In line with the prevailing political environment Powell declared the mound builders did not have any connections with the Old World, despite the fact that his own archaeologists had discovered the Bat Creek Stone in an official dig.

According to Scherz, the publication of Powell's conclusions allowed the American army to continue driving Indians from their lands "without guilt they might feel if they thought themselves somehow related to the refugees" (pp. 1-5). The Indian lands were taken, the wars ended, but "the influence of Powell is with us still in our universities, public institutions, libraries and book stores" (pp. 1-5).

The methods used by Powell and others in response to opposition to their archaeological dogmas warrant careful scrutiny. For example, Hjalmar Holand struggled for forty years to have the authenticity of the Kensington Stone accepted. Throughout this entire time unrelenting pressures and persecutions were heaped upon him. The result of this type of action "has completely distorted history of our country merely to save face and uphold an ipse dixit pronouncement laid down many years ago by an academic group intrusted by the public with preservation of these matters" (Mertz, 2004, p. 67).

"Ipsi dixit" is a little used term that accurately describes the actions of Powell and others in relation to the mound builder issue.

> (Latin for "he said it himself") is an assertion without proof; or a dogmatic expression of opinion. The fallacy of defending a proposition by baldly asserting that it is "just how it is" distorts the argument by opting out of it entirely: the claimant declares an issue to be intrinsic, and not changeable (Ipsi dixit, n.d.).

Silverberg (1970) commented on a dispute between the Smithsonian Bureau of Ethnology and the Davenport Academy of Natural Sciences. Putnam, from the Davenport Academy, feared that the Smithsonian was using its power and influence "to impose an intellectual tyranny on American archaeology" (p. 143). Its use of scorn and suppression stifled independent thought. The Smithsonian was victorious over the Academy and according to Silverberg (1970) "in decades to come, on other subjects, the archaeologists who spoke for the Smithsonian were indeed able to impose their ideas on others-and they were not always correct" (p. 143).

It seems that Scherz was right, "Powell is with us still in our universities, public institutions, libraries and book stores" (Mertz, 2004, p. x).

But what about the Church of Jesus Christ of Latter-Day Saints, why did it promote Central and South America and why do Mayan temples still adorn artworks hung in Church properties? The mechanics of the initial deception were well explained by Jonathan Neville (n.d.) in "The Smoking Gun of Book of Mormon Geography". According to Neville, in 1842 two men, Benjamin Winchester and William Smith conspired to deceive members of the Church into believing that Central America was the setting for the Book of Mormon.

Benjamin Winchester (1817-1901)

In 1841 John L. Stephens published a book titled *Incidents of Travel in Central America, Chiapas and Yucatan*. The book became a best seller and Winchester believed that he would persuade more people to read the Book of Mormon and join the Church if he linked it to the sensational best seller. With the help of William Smith, who later apostatized, spurious articles were published in *The Times and Seasons*. At the time of the publication, in 1842, Joseph Smith was in hiding. When he became aware of the articles, he released Winchester from his Church calling and fired William Smith from his editorial position with another Church paper, 'The Wasp'.

In 1844 Joseph Smith was martyred. Soon after the Church members began leaving their homes for the long and dangerous trek to settle in the valley of the Great Salt Lake. In the midst of all this turmoil it seems that the deceptions initiated by Benjamin Winchester and William Smith went uncorrected.

A little over a century later, in 1946, Brigham Young University began offering courses in Archaeology. Some of its leading academics focused their studies on central America and attempted to establish ties between central American archaeology and the Book of Mormon. The further they pursued this course the more committed they became to central America as the location for the Book of Mormon. Findings that did not match the Book sometimes required novel solutions, hence, as previously discussed, the lamb of God became the agouti of God. They built their careers on this linkage.

At around the same time Arnold Friberg was commissioned to provide illustrations of Book of Mormon scenes. His drawings completed the mental images of the Book of Mormon for many members of the Church. Unfortunately, they resembled Mayan ruins from central America and ignored the text of the Book itself. Eventually we had lesson manuals, teaching aids, videos and paintings adorning our chapel walls that featured Mayan temples.

There was obviously a lot more to the story but in the end, people accepted what was being presented and the theory became fact. For a time, the question of geography didn't matter that much, but the situation changed when critics started presenting seemingly compelling reasons to refute the link with Central America. That did matter, breaking that linkage was an attack on the cornerstone of the religion. For some members it created a crisis of faith.

7

The Cultural Case

THIS RESEARCH WAS INITIALLY COMMENCED to confirm or dismiss the arguments for the North American theory put forward by people like Rod Meldrum. The process involved reading books and papers written by experts in the fields of archaeology, geology and other sciences who had no links to the Church of Jesus Christ of Latter-Day Saints. In the course of reading it became clear that in addition to the arguments already put forward, there were many correlations between the cultural evidences in North America and the text of the Book of Mormon.

The Archaic Period Cultures were included for completeness. There is simply not enough information available to draw significant conclusions but there is in the Copper Culture a candidate for the Jaredites nation of the Book of Mormon.

It was located in the land to the north of the proposed Nephite lands. That land was within reach of the Hill Cumorah. It was near seas which "divide the land" and narrow necks of land. It was a land of many waters. It was a land endowed with all the minerals and ores described in the Book of Mormon. The Copper Culture, like the Jaredites, was involved in significant mining. It existed during the relevant time period and disappeared suddenly. It also provided speculation as to how the Mulekites later travelled from the Middle-east and arrived where they did, in the Jaredite heartland.

The Early Woodland Period commenced with an uninhabited landscape. The timing of its appearance fits in well enough with the arrival of the Lehites when you consider the small numbers of people involved. Markers of the Early Woodland Culture diffused from the Southeast in a westerly and northerly direction. The settlement patterns of communities initially in the upland areas and more primitive lifestyles in the west accord with the Book of Mormon.

The Adena Culture shared similarities with the Mulekites of the Book of Mormon. The timing of their appearance and the fact that the culture was not the result of local evolution but of intrusion is the same. They arrived in their territory before the Hopewell just as the Mulekites settled their lands and were later joined by the Nephites. When the Hopewell appeared, they co-existed with and dominated the Adena despite the Adena being more numerous, just as the Nephites, though fewer in number, dominated the Mulekites.

The date of the Hopewell Culture's arrival in the Illinois River valley matches the arrival of the Nephites in Zarahemla. The date of their spread into the Ohio River valley matches the Nephite movement into other areas. The order in which the settlement took place agrees.

The Hopewell, like the Nephites, grew crops. Both group's diets included corn and barley, the latter being a staple food.

The Hopewell and the Nephites both wore pearls and both worked in different types of metals. They also both constructed extensive road networks.

The Hopewell territories included unique geological features. They encompass the area in which the largest earthquake in United States history occurred as did prehistoric earthquakes. They are also affected by isostatic rebound. These forces have combined to produce effects such as upward and downward land movements, the creation of new lakes and domes, land fissuring, quaking, and the mass ejection of materials into the air causing darkness. It is also an area subject to tornadoes. All of these phenomena are described in the Book of Mormon.

The Hopewell were religious and displayed no signs of idol worship or polytheism. They built sacred enclosures with architectural symbolism that parallels symbolism in modern architecture of the Church of Jesus Christ of Latter-Day Saints. These enclosures contained lunar alignments, enabling them to comply with the law of Moses. They also built temple mounds without steps, as required by the Mosaic law. The Hopewell took their religion to the Adena, just like the Nephites took theirs to the Mulekites.

Shortly before the birth of Christ the Hopewell began trading with their southern neighbours. For a little over two centuries goods were exchanged with the people of the Marksville Culture. Then as suddenly as it began the trade ended, precisely as trade between the Nephites and Lamanites is described in the Book of Mormon.

During the period from the first half of the first century A.D. till 200 A.D. the Hopewell experienced a renaissance. Their culture spread throughout the Midwest and brought changes to technology, subsistence, community, ideology and settlement patterns. These changes mirrored the golden age of the Nephites when, following the visit of the resurrected Saviour, they experienced a period of unity with their neighbours and unprecedented prosperity.

Hopewell architecture was based on the use of timber, just as the Book of Mormon makes clear that Nephite architecture was based on timber. They fortified their settlements in precisely the same way the Hopewell fortified theirs including earthen walls topped with palisades and flanked by an exterior ditch. They also both built places of retreat with unusual places of entry. These structures formed part of a system of defences designed to protect them from their southern neighbours with whom they were engaged in a protracted struggle.

At some time following 300 A.D. the Hopewell began to withdraw from their prime lands into larger, fortified communities. They ceased building ceremonial centres and burial mounds. In around 350 A.D. they abandoned their lands in the Mississippi Valley. They also constructed a cordon of heavy defences in the north, just below the Great Lakes. In around 400 A.D. the Hopewell Culture disappeared from the land. All of this concurs with the Book of Mormon.

The Hopewell have been found in mass graves. They used armour like that used by the Nephites. There is evidence that they made weapons of metal. They wrote in hieroglyphics and Hebrew characters. They have much in common with the Nephites

The North American theory resolves a number of problems associated with the Central American theory. The Hill Cumorah is where it should be. North is where it should be. The animals and plants required to live the law of Moses are where they should be. The prophecies in 2 Nephi are where they should be. The geological and climatological features are where they should be. Haplogroup X DNA is where it should be.

The Book of Mormon text is full of cultural features and dates that align perfectly with the archaeology of North America. How is this possible? Joseph Smith grew up in that area at a time when the population was debating the mound builders, but most of the similarities outlined in this text were not known in the nineteenth century. The relationship between the Hopewell, the Adena and the Marksville, the timing of their trade and of the golden age were all yet to be discovered. There are too many details and too many dates which we now know match to be coincidence.

A FRESH PERSPECTIVE

The following points are offered to help the reader adjust their mental images as they read the Book of Mormon.

NEW FILTER

Adjust your image of what the Nephites looked like. Forget the dark hair and swarthy complexion of Friberg artworks. According to Nephi the Nephites looked just like the Gentiles that would later inhabit the land.

> And I beheld the Spirit of the Lord, that it was upon the Gentiles, and they did prosper and obtain the land for their inheritance; and I beheld that they were white, and exceedingly fair and beautiful, like unto my people before they were slain (1 Nephi 13:15).

LEHITE JOURNEY TO AMERICA

Following decades of research by people including Hope and Lynn Hilton and Warren and Michaela Aston, many members of the Church agree that the Lehites travelled from Jerusalem to Oman where they began their sea journey to the Americas. Despite there being little evidence in the text to indicate the sailing route a Pacific crossing to the west coast of America has generally been accepted. This notion should be challenged.

The Book of Mormon describes the beginning of the sea voyage as follows:

> 1 Nephi 18:6 ... after we had prepared all things, much fruits and meat from the wilderness, and honey in abundance, and provisions according to that which the Lord had commanded

us, we did go down into the ship, with all our loading and our seed ...

1 Nephi 18:8 ... after we had all gone down into the ship ... we did put forth into the sea and were driven forth before the wind ...

The Lehites completed their preparations, gathered their provisions of fruit and meat and honey, loaded their vessel and set sail, being driven before the wind. The text does not reveal what time of year this occurred in or in what direction they sailed so we must look elsewhere for clues.

The winds in the Gulf of Oman and the Northern Arabian Sea are dominated by summer and winter monsoons. From November to April the prevailing winds are from the north-east, from July to September persistent southerly and south-westerly winds occur, and during the months in between the wind patterns are less persistent (Chaichitehrani & Allahdadi, 2018). For the Lehites to take advantage of prevailing winds they would have departed in the period from November to April and sailed in a south-westerly direction, along the eastern coastline of Africa. This timing would have the added advantage of avoiding the oppressive summer temperatures.

The viability of such a journey was demonstrated by the Phoenician Ship Expedition. Philip Beale built a replica Phoenician ship, a type of ship that was in use around 600 B.C. and attempted to circumnavigate Africa. The ship left Oman on 26 October 2009 and sailed down the east coast of Africa. It rounded the Cape of Good Hope and attempted to sail northward, along the west coast of Africa but, despite their best efforts, the prevailing trade winds and ocean currents carried the ship across the Atlantic to within a few hundred miles of making landfall in North America. They continued northward till they picked up the Gulf Stream and recrossed the Atlantic before sailing south and into the Mediterranean. This journey demonstrated the that a ship of the type used in 600 B.C. could sail from Oman to America across the Atlantic Ocean.

The Book of Mormon describes a significant weather event that occurred during the Lehite voyage.

1 Nephi 13:13 ... there arose a great storm, yea, a great and terrible tempest, and we were driven back upon the waters for the

space of three days; and they began to be frightened exceedingly lest they should be drowned in the sea ...

14 And on the fourth day, which we had been driven back, the tempest began to be exceedingly sore.

15 And it came to pass that we were about to be swallowed up in the depths of the sea ...

The Cape of Good Hope was previously named the Cape of Storms. It is recognised as one of the most dangerous places in the world to sail because of the violence of its storms with waves over five metres and winds exceeding 30 knots, freak waves and unpredictable cross currents. Could the storm described in the Book of Mormon been one of these storms?

There is enough evidence to prove that the Lehite journey from Oman to the Americas could have been across the Atlantic Ocean. This is a considerably shorter and quicker route. Perhaps the voyage took advantage of winds and currents to cross the Atlantic and arrive on the southern coast of the USA.

LEXICAL DIFFERENCES

Be careful with terminology. If you study different languages you will find that they do not always have lexically equivalent words, that is, a word in one language may not have a matching word in another language with precisely the same meaning. One example that I mentioned earlier is the word "sea". The Israelites were not familiar with oceans so any body of water bigger than a pond was called a sea. The Bible refers to the sea of Galilee, a body of water approximately 21 kilometres by 13 kilometres in size. The Book of Mormon always uses different terms such as "many waters" or "great waters" when referring to oceans, so when you read the word sea don't think ocean, think water surrounded by land.

One of the most persistent problems with proposed maps of Book of Mormon geography is that they are generally flanked on the left and on the right by oceans. When they add a narrow neck, it produces a shape that resembles Central America. However, when Hagoth launched his ship "on the borders of the land Bountiful, by the land Desolation, and launched it forth into the west sea, by the narrow neck which led into the land northward (Alma 63:5)" and sailed northward, it could not be referring to the Pacific Ocean coast of Central America. It is not possible to launch into the Pacific Ocean from Central America and sail north as the coastline runs in a north-westerly direction and Central America is in the way. It is possible to launch into the Great Lakes, by a narrow neck of land leading into the land northward and sail north.

Take time to understand unfamiliar terms. For example, what does "head of the river" mean? The Collins English Dictionary defines River Head as "the source of a river." In many cases a river has multiple sources or heads. The Mississippi is one such river with multiple major heads including the Ohio River, Illinois River, Missouri River and Mississippi River. When the Book of Mormon repeatedly refers to the River Sidon and only the River Sidon is it because to the Nephites there was only one major River in their territories, albeit with multiple heads? When Alma wrote "by the head of the river Sidon, running from the east towards the west" (Alma 22:27) was he referring to a head or tributary of a river that could be traced from its confluence in an east to west direction?

PUNCTUATION

Be careful of punctuation. "Reformed Egyptian" has no punctuation and when Joseph Smith translated the Book of Mormon it was dictated and written without it. Punctuation was first added by John Gilbert, the typesetter and compositor at the printers.

There are a few very wordy sentences where punctuation can be confusing and it is easy to miss the meaning. The best example is Alma 22:27.

> And it came to pass that the king sent a proclamation throughout all the land, amongst all his people who were in all his land, who were in all the regions round about, which was bordering even to the sea, on the east and on the west, and which was divided from the land of Zarahemla by a narrow strip of wilderness, which ran from the sea east even to the sea west, and round about on the borders of the seashore, and the borders of the wilderness which was on the north by the land of Zarahemla, through the borders of Manti, by the head of the river Sidon, running from the east towards the west—and thus were the Lamanites and the Nephites divided.

That is a very complex sentence. The difficulty is working out what is the correct subject that each of the dependent clauses refer to. My suggestion is as follows:

> And it came to pass that the king sent a proclamation throughout all the land, amongst all his people who were in all his land, who were in all the regions round about,

The subject—the king's land and the regions round about
Dependent clauses relating to the king's land (and the regions round about)—

> which was bordering even to the sea, on the east and on the west,
> and which was divided from the land of Zarahemla by a narrow strip of wilderness,
> which ran from the sea east even to the sea west,
> and round about on the borders of the seashore,
> and the borders of the wilderness

We could read each clause in conjunction with the subject like this:

> The king's land was bordering even to the sea on the east and on the west
> The king's land was divided from the land of Zarahemla by a narrow strip of wilderness
> The king's land ran from the sea east even to the sea west
> The king's land ran round about on the borders of the seashore
> The king's land ran round about on the borders of the wilderness

The second subject—the wilderness
Dependent clauses relating to the wilderness

> which was on the north by the land of Zarahemla,
> through the borders of Manti,
> by the head of the river Sidon,

We could read each clause in conjunction with the second subject like this:

> The wilderness was on the north by the land of Zarahemla,
> The wilderness ran through the borders of Manti,
> The wilderness ran by the head of the river Sidon,

The third subject—the head of the river Sidon
Dependent clause relating to the river Sidon

> running from the east towards the west

We could read this clause in conjunction with the third subject like this:

The head of the river Sidon running from the east towards the west

The sentence then concludes with—and thus were the Lamanites and the Nephites divided.

NAMES VS DESCRIPTIONS

Some mentions of physical features are names of individual landmarks and others are not. So, the sea, on the west is a description of where the sea is located in relation to other landmarks rather than the name of the sea.

MODERN MAPS

Expect to find some differences between modern maps and the text of the Book of Mormon. The landscape has changed. The Book of Mormon itself testifies to this:

> 3 Nephi 8:12 ... the whole face of the land was changed, because of the tempest and the whirlwinds, and the thunderings and the lightnings, and the exceedingly great quaking of the whole earth;
>
> 13 And the highways were broken up, and the level roads were spoiled, and many smooth places became rough.
>
> 14 And many great and notable cities were sunk, and many were burned, and many were shaken till the buildings thereof had fallen to the earth ...
>
> 17 And thus the face of the whole earth became deformed, because of the tempests, and the thunderings, and the lightnings, and the quaking of the earth.
>
> 18 And behold, the rocks were rent in twain; they were broken up upon the face of the whole earth, insomuch that they were found in broken fragments, and in seams and in cracks, upon all the face of the land.

The earthquake that caused much of this destruction was discussed earlier. If the Midwest and particularly the area around the Great Lakes is associated with the Book of Mormon, we should be cognizant that the landscape has undergone many natural changes in relatively recent times.

It has also undergone man made changes. For example, a swamp approximately 100 miles long and 25 miles wide extended from the western edge of Lake Erie, greatly narrowing the passage into Michigan. This swamp was drained in the late 1800s. Also, the Mississippi River has changed. Apart from naturally occurring changes as old channels silt and new ones open many dams and causeways have been constructed. Even before this the practice of cutting down trees to fuel steamboats resulted in the river becoming wider and shallower.

The Book of Mormon states that the changes at the time of the Saviour's death caused the residents of Bountiful to be "marveling and wondering one with another, and were showing one to another the great and marvelous change which had taken place". This confirms that significant changes occurred in ancient times. We should expect some differences between the text of the Book of Mormon and the landscape we see today. (3 Nephi 11:1).

POPULATION SIZES

Be careful not to overestimate the size of populations in the Book of Mormon. In about 120 B.C. Alma established churches throughout the land of Zarahemla because there were too many people to be governed by one teacher (Mosiah 25:20). The total number of churches (or congregations) established was seven. 500 years later Mormon wrote that the number of Nephite deaths in the final battle at Cumorah was 230,000. There was also an unknown number of defectors.

TRUST THE SCRIPTURES.

Ask yourself why the Doctrine and Covenants asks the Saints to build a city named Zarahemla on the western side of the Mississippi River, opposite Nauvoo.

TRUST JOSEPH SMITH

He was the most knowledgeable and qualified man in this dispensation to speak on the Book of Mormon and his statements and actions should be accorded the appropriate respect.

Reference List

1811-12 New Madrid earthquakes. (n.d.) In *Wikipedia*. Retrieved October 16, 2016, from http://en.wikipedia.org/wiki/1811%E2%80%9312 _New_Madrid_earthquakes

Abrams, E. (2009). Hopewell Archaeology: A View from the Northern Woodlands. *Journal of Archaeological Research, 17*(2), 169-204.

Abrams, E., & Freter, A. (Eds.) (2005). *The Emergence of the Moundbuilders: The Archaeology of Tribal Societies in Southeastern Ohio*, Ohio University Press, Athens.

Anderson, D. G. (2008). The End of the Southeastern Archaic: Regional Interaction and Archaeological Interpretation (pp. 273-298). In D. H. Thomas and M. C. Sanger (Eds). *Trend, Tradition, and Turmoil. What happened to the Southeastern Archaic?* Proceedings of the Third Caldwell Conference, St. Catherine's Island, Georgia, May 9-11, 2008.

Archaeology Wordsmith. (2002-2018). *Tradition.* Retrieved from https:// archaeologywordsmith.com on 20/11/2018.

Bat Creek Inscription. In *Wikipedia*. Retrieved 31/12/2018 from https:// en.wikipedia.org/wiki/Bat_Creek_inscription

Barkay, G. (2003). Mounds of mystery: where the kings of Judah were lamented. *Biblical Archaeology Review, 29* (3), pp. 32–9, 66, 68.

Birmingham, R. A., & Eisenberg, L. E. (2000). *Indian Mounds of Wisconsin.* Madison, WIS: The University of Wisconsin Press.

Book of Mormon and DNA Studies. (2014). Retrieved from https://www.LDS.org/manual/gospel-topics/ book-of-mormon-and-DNA-studies?lang=eng

Boudinot, E. (1816). *A Star in the West.* Trenton, N.J.: George Sherman.

Brown, J. A. (1979). Charnel Houses and Mortuary Crypts: Disposal of the Dead in the Middle Woodland Period (Chapter 27). In D. S. Brose & N. Greber (Eds.). *Hopewell Archaeology. The Chillicothe Conference.* Kent, Ohio: Kent State University Press.

Butler, B. M. (1979) Hopewellian Contacts in Southern Middle Tennessee. In D. S. Brose and N. Greber, (Eds). Hopewell Archaeology: The

Chillicothe Conference, *Midcontinental Journal of Archaeology, Special Papers*, 3. pp. 150-156.

Carskadden, J., & Morton, J. (1997). Living on the Edge: A Comparison of Adena and Hopewell Communities in the Central Muskingum Valley of Eastern Ohio (Chapter 14). In W. S. Dancey & P. J. Pacheco (Eds.). *Ohio Hopewell Community Organization.* Kent, Ohio: Kent State University Press.

Chaichitehrani, N., & Allahdadi, M. N. (2018). Overview of Wind Climatology for the Gulf of Oman and the Northern Arabian Sea. *American Journal of Fluid Dynamics*, 8(1), pp. 1-9.

Charles River Editors (2015). Native American tribes: The history and culture of The Mound Builders. Harvard and MIT Alumni.

Cullen, K. M. (2006). *Old Copper Culture.* Wi, USA: Milwaukee Public Museum. Retrieved from https://www.mpm.edu/research-collections/ anthropology/online-collections-research/old-copper-culture on 2/12/2018.

Dancey, W. S., & Pacheco, P. J. (1997). *Ohio Hopewell Community Organization.* Kent, Ohio: Kent State University Press.

DNA vs. The Book of Mormon (2003). Living Hope Ministries/sourceFLIX.

Cannon, D. Q. (1995). Zelph Revisited, In Garrett, H. D. (Ed.), *Church History Regional Studies, BYU Department of Church History and Doctrine, Regional Studies, Illinois*, pp. 97-109. Retrieved 2/12/2018 from http:// www.bhporter.com/Porter%20PDF%20Files/ZelphRevisitedCannon[1]. pdf

Fortier, A. (2001). A Tradition of Discontinuity (Chapter 11). In T. R. Pauketat (Ed.) *The Archaeology of Traditions.* Gainsville, Florida: University Press of Florida.

Gibson, J. L. (2008). "Nothing but the river's flood": Late Archaic diaspora or disengagement in the lower Mississippi Valley and southeastern North America (pp. 33-44). In D. H. Thomas and M. C. Sanger (Eds). *Trend, Tradition, and Turmoil. What happened to the Southeastern Archaic?* Proceedings of the Third Caldwell Conference, St. Catherine's Island, Georgia, May 9-11, 2008.

Grena, G. M. (2004). *LMLK: A Mystery Belonging to the King vol. 1.* Redondo Beach, California: 4000 Years of Writing History.

Hoffman, D. (2014). *Missing: Prehistoric Michigan's Half-Billion Pounds of Copper.* Ancient America. Retrieved from http://ancientamerica.

com/missing-prehistoric-michigans-half-billion-pounds-of-copper/ on 30/11/2018.

Ipsi dixit. (n.d.) in *Wikipedia*. Retrieved 28/12/2018 from https://en.wikipedia.org/wiki/Ipse_dixit

Joseph Fielding Smith (1956), *Doctrines of Salvation*, edited by Bruce R. McConkie. Salt Lake City: Bookcraft.

Kozarek, S. E. (1997). Determining sedentism (Chapter 5). In W. S. Dancey & P. J. Pacheco (Eds.). *Ohio Hopewell Community Organization*. Kent, Ohio: Kent State University Press.

Kidder, T. R. (2008). Trend, tradition and transition at the end of the Archaic (pp. 23-32). In D. H. Thomas and M. C. Sanger (Eds). *Trend, Tradition, and Turmoil. What happened to the Southeastern Archaic?* Proceedings of the Third Caldwell Conference, St. Catherine's Island, Georgia, May 9-11, 2008.

Kingsley, R. G. (1981). Hopewell Middle Woodland Settlement Systems and Cultural Dynamics in Southern Michigan. *Midcontinental Journal of Archaeology*, Vol. 6, No. 2. Maney Publishing on behalf of the Midwest Archaeological Conference, Inc.

Lankford, G. E. (2007). *Reachable stars: Patterns in the ethnoastronomy of eastern North America*. University of Alabama Press.

Mallory, A., & Roberts-Harrison, M. (1979). *The Rediscovery of Lost America*. New York: E. P. Dutton.

Manifest Destiny. (n.d.). In *Wikipedia*. Retrieved 28/12/2018 from https://en.wikipedia.org/wiki/Manifest_destiny

Marder, W. (2005). *Indians in the Americas. The Untold Story*. San Diego, CA: The Book Tree.

Meldrum, R. L. (2011). *Exploring the Book of Mormon in America's Heartland. A Visual Journey of Discovery*. New York: Digital Legend Press and Publishing.

Mertz, H. (2004). *The Mystic Symbol. Mark of the Michigan Mound Builders*. Colfax, Wisconsin: Hay River Press.

Miller, S., & Lipsett, S. (1998). *After death: How people around the world map the journey after life*. New York, NY: Simon and Schuster.

Mock, R. D. (2007). *Passover Seder*. Downloaded 2/5/2016 from http://www.biblesearchers.com/hebrews/festivals/passover.shtml

Monroe Doctrine. (n.d.). In *Wikipedia*. Retrieved 28/12/2018 from https://en.wikipedia.org/wiki/Monroe_Doctrine

Nelson, A. (2012). *5 Scriptures that Validate the Heartland Model*. DVD.

Neuman, R. W., & Hawkins, N. W. (1993). *Louisiana Prehistory* (Second Edition). LA, USA: Louisiana Archaeological Survey and Antiquities Commission, Department of Culture, Recreation and Tourism.

Neville, J. (n.d.). *The Smoking Gun of Book of Mormon Geography*. Retrieved 28/12/2018 from http://firmlds.org/smoking-gun-book-mormon-geography/

Poverty Point Culture. (n.d.). In *Wikipedia*. Retrieved November 27, 2018, from http://en.wikipedia.org/wiki/Poverty_Point_culture

Poverty Point Culture. (2013). *The Concise Oxford Dictionary of Archaeology* (2). Retrieved from www.oxfordreference.com

Prufer, O. H. (1997). Forthill 1964, (Chapter 12). In W. S. Dancey & P. J. Pacheco (Eds.). *Ohio Hopewell Community Organization*. Kent, Ohio: Kent State University Press.

Prufer, O. H. (1964). *The Hopewell Cult*. Scientific American, Vol. 211, No. 6.

Raff, J. A., & Bolnick, D. A. (2015). Does Mitochondrial Haplogroup X Indicate Ancient Trans-Atlantic Migration to the Americas? A Critical Re-Evaluation. *PaleoAmerica, 1*, (4), pp. 297-304. https://doi.org/10.1179/2055556315Z.00000000040.

Randall, E. O. (1908). *The masterpieces of the Ohio mound builders. The hilltop fortifications*. Columbus, OH: Ohio State Archaeological and Historical Society.

Reed, S. (2011). *Discovering sacred teachings in LDS Chapel architecture*. Downloaded 1/5/2016 from http://oneclimbs.com/2011/11/14/discovering-sacred-teachings-in-lds-chapel-architecture/

Rich, T. R. (n.d.). *Jewish Calendar*. Retrieved from http://www.jewfaq.org/calendar.htm on 31/3/2018.

Riley, T., Edging, R., & Rossen, J. (1990). *Cultigens in Prehistoric North America. Changing Paradigms*. Current Anthropology. 31(5):525-541.

Romain, W. F. (2015). *An Archaeology of the Sacred—Adena-Hopewell Astronomy and Landscape Archaeology*. Olmsted Township, Ohio: The Ancient Earthworks Project.

REFERENCE LIST

Romain, W. F. (2000). *Mysteries of the Hopewell. Astronomers, Geometers, and Magicians of the Eastern Woodlands.* Akron, Ohio: The University of Akron Press.

Rusch, E. (2011). *The Great Midwest Earthquake of 1811.* In Smithsonian Magazine. Retrieved 21 December, 2018, from http://www.smithsonianmag.com/science-nature/the-great-midwest-earthquake-of-1811-46342/#cFVJcLUV5XYAJSlq.99

Scherz, J. P. (2001). *America's Pre-Columbian Monroe Doctrine.* Madison, Wisconsin. University of Wisconsin.

Schweikart, J. F. (2008). 8 Upland Settlement in the Adena Heartland. *Transitions: Archaic and Early Woodland Research in the Ohio Country,* 183.

Silverberg, R. (1970). *The Mound Builders,* OH: Ohio University Press.

Smith, B. A. (1979). The Hopewell Connection in Southwest Georgia (Chapter 24). In D. S. Brose & N. Greber (Eds.). *Hopewell Archaeology: The Chillicothe Conference.* Kent, Ohio: Kent State University Press.

Smith, L. M. (1853). Joseph Smith the Prophet: Biographical sketches of Joseph Smith the Prophet and his progenitors for many generations. Liverpool, UK: S.W. Richards, 15 Wilton Street, Liverpool.

Sorenson, J. L. (1985). *An ancient American setting for the Book of Mormon.* Deseret Book Company.

Spaulding, A. C. (1952). *The Origin of the Adena Culture of the Ohio Valley,* Southwestern Journal of Anthropology, Museum of Anthropology, University of Michigan, Michigan.

Squier, E. G., & Davis, E. H. (1988). *Ancient Monuments of the Mississippi Valley.* Smithsonian Institution.

Stover, C. W., & Coffman, J.L. (1989) *Seismicity of the United States, 1568-1989 (Revised),* US Geological Survey Professional Paper 1527, United States Government Printing Office, Washington.

Strange Happenings during the Earthquakes, (n.d.) New Madrid. Retrieved December 12, 2018 from http://www.new-madrid.mo.us/index.aspx?nid=132

Toth, A. (1979). The Marksville Connection (Chapter 25). In D. S. Brose & N. Greber (Eds.). *Hopewell Archaeology The Chillicothe Conference.* Kent, Ohio: Kent State University Press.

Tumulus. (n.d.). in *Wikipedia.* Retrieved 18 December, 2018, from https://en.wikipedia.org/wiki/Tumulus#Israel

Wagenaar, J. A. (2005). Origin and Transformation of the Ancient Israelite Festival Calendar (Vol. 6). Otto Harrassowitz Verlag.

Wakefield, J. S. (2015). *Were Prehistoric Copper Oxhide Ingots manufactured on the Mississippi coast near the mouth of the Mississippi River.* Ancient America. Retrieved from http://ancientamerica.com/oxhides/#more-5533 on 30/11/2018.

Woodland Period. (n.d.). in *Wikipedia*. Retrieved 9 December, 2018, from https://en.wikipedia.org/wiki/Woodland_period

Woodward, S., & McDonald, J. (1986). *Indian Mounds of the Middle Ohio Valley. A guide to Adena and Ohio Hopewell Sites.* Newark, OH: The McDonald & Woodward Publishing Company.

Woodward, S., & McDonald, J. (2001). *Indian Mounds of the Middle Ohio Valley. A guide to mounds and earthworks of the Adena, Hopewell, Cole, and Fort Ancient people.* Newark, OH: The McDonald & Woodward Publishing Company.

Wright, H. E., & Frey, D. G. (2015). *The Quarternary of the US.* Princeton University Press.

Index

Made in the USA
Monee, IL
14 March 2026

46140448R00073